T0215309

SCIENCE AND MATHEMATICS

This book offers an engaging and comprehensive introduction to scientific theories and the evolution of science and mathematics through the centuries. It discusses the history of scientific thought and ideas and the intricate dynamic between new scientific discoveries, scientists, culture and societies.

Through stories and historical accounts, the volume illustrates the human engagement and preoccupation with science and the interpretation of natural phenomena. It highlights key scientific breakthroughs from the ancient to later ages, giving us accounts of the work of ancient Greek and Indian mathematicians and astronomers, as well as of the work of modern scientists like Descartes, Newton, Planck, Mendel and many more. The author also discusses the vast advancements which have been made in the exploration of space, matter and genetics and their relevance in the advancement of the scientific tradition. He provides great insights into the process of scientific experimentation and the relationship between science and mathematics. He also shares amusing anecdotes of scientists and their interactions with the world around them.

Detailed and accessible, this book will be of great interest to students and researchers of science, mathematics, the philosophy of science, science and technology studies and history. It will also be useful for general readers who are interested in the history of scientific discoveries and ideas.

Jayant V. Narlikar spent most of his learning life at Cambridge, England, including a Ph.D. followed by Sc.D. He was a Fellow of King's College. Besides a highly productive career as a professional astrophysicist, he has acquired the reputation of someone who has popularized science and as an author of science fiction for which he was awarded the prestigious Kalinga award by UNESCO.

SCIENCE AND MATHEMATICS

From Primitive to Modern Times

Jayant V. Narlikar

LONDON AND NEW YORK

First published 2022
by Routledge
2 Park Square, Milton Park, Abingdon, Oxon OX14 4RN

and by Routledge
605 Third Avenue, New York, NY 10158

Routledge is an imprint of the Taylor & Francis Group, an informa business

© 2022 Jayant V. Narlikar

The right of Jayant V. Narlikar to be identified as author of this work
has been asserted by him in accordance with sections 77 and 78 of the
Copyright, Designs and Patents Act 1988.

All rights reserved. No part of this book may be reprinted or reproduced or
utilised in any form or by any electronic, mechanical, or other means, now
known or hereafter invented, including photocopying and recording, or in
any information storage or retrieval system, without permission in writing
from the publishers.

Trademark notice: Product or corporate names may be trademarks or
registered trademarks, and are used only for identification and explanation
without intent to infringe.

British Library Cataloguing-in-Publication Data
A catalogue record for this book is available from the British Library

Library of Congress Cataloging-in-Publication Data
A catalog record for this book has been requested

ISBN: 978-0-367-64005-7 (hbk)
ISBN: 978-1-032-06624-0 (pbk)
ISBN: 978-1-003-20310-0 (ebk)

DOI: 10.4324/9781003203100

Typeset in Bembo
by Apex CoVantage, LLC

To
the late Jean-Claude Pecker
from whom I learnt to enjoy
and criticize science

CONTENTS

FIGURES

PREFACE

A few years ago, I wrote a book in *Marathi*, the written and spoken language of the western part of India called Maharashtra (literally meaning 'big nation.'). The book contained a brief history of how science and mathematics have grown as human civilization progressed. The growth has sometimes been independent, while there have been occasions when one inspired the other. My attempt at demonstrating a relationship between them received a good reader response. So much so that the book was awarded a Maharashtra government prize in the category of science for a general readership.

This recognition encouraged me to write an English version of the aforementioned book with suitable modifications keeping in mind the worldwide general readership. Also, I took this opportunity to add extra topics so as to make the various discussions deeper.

Perhaps I should clarify that this is not a comprehensive historical account like the classic book *Men of Mathematics* by E. T. Bell or an extensive collection of scholarly articles like James R. Newman's *The World of Mathematics*. The mode used here is more like that of storytelling and I hope it succeeds.

To correct a possible impression that scientists and mathematicians are extra-superior persons, I have given at the end a collection of anecdotes about them, showing that apart from their own particular fields, they are no different than ordinary people like us. The anecdotes described are indicative only and not all of them are to be taken literally as true. Many of them have been in circulation in person-to-person mode and not all of them could be identified with specific sources.

In Indian classical music, there are occasions when two artists playing different musical instruments get into a competitive mood. Thus, we may have one artist playing the *Sitar* and another playing the harmonium. In the competitive mode, they try to surpass each other. There are criteria which tell who amongst them is "winning" the duet competition. Whosoever may be the formal winner, the musical piece itself rises in quality. The conclusion in the epilogue conveys a similar impression with mathematics and science playing the duet.

It is a pleasure to thank all those whose advice and help made it possible to prepare this book. In particular, I want to mention Simon Mitton whose advice certainly helped. Mangala, my wife read the manuscript and gave several comments both as a lay reader and as a professional mathematician. I thank Prem Kumar, Suhas Jagtap and Vishwas Kale for preparing the illustrations and images. And last but not least, I would like to thank my secretary Vyankatesh Samak for helping put finishing touches on the manuscript. Barring a few (where contribution is acknowledged) most images used are taken from the public domain on the web. Finally, it has been a pleasure to interact with the editorial teams of the publishers.

Jayant Narlikar

Inter-University Centre for Astronomy and Astrophysics,

Pune University Campus,

Pune 411007

PROLOGUE

The story of mathematics and science

This is the story of how the subjects of mathematics and science grew from a very primitive state to the sophisticated levels at which we find them today. The early history that has been pieced together tells us that in primitive times, the humans lived in caves and managed to subsist on food that was vegetarian like fruits on trees or non-vegetarian like birds and animals that were hunted down.

Imagine a family in those times. An apple tree in their neighbourhood produced a lot of fruit. The ripe ones were taken down and stored by the head of the family. When the kids started clamouring for them, he made them sit down in a row and started distributing them. He found at the end of the exercise that some apples were left. If he knew how to count, he might have discovered that he had 25 apples to start with and the number of kids was 7. And at the end of the exercise, everyone would have gotten three apples each, and four would be left.

The caveman repeated this experiment with other fruit when they were ripe and ready to eat. He got the same answer each time. And if he were attentive, he would have discovered the rules of addition, multiplication: $3 \times 7 + 4 = 25$.

Experiences in day-to-day living would thus have brought mathematics as a useful tool. As the example shows, a clever caveman could have learnt the elementary mathematical operations *without* learning to write. The system of writing numbers and using the operations as mathematical statements came because of the need for it.

Side by side, the caveman also observed his environment and began to discover some patterns and periodicities. A major discovery may have been that of fire and learning how to produce it. Special stones rubbed suitably could produce fire – the same fire that he may have encountered in dry forest areas. Just as natural fire could spread by 'swallowing,' he could also generate his fire by feeding suitable material. Compared to uncontrolled natural fire, his locally generated fire could be put to a lot of use. Cave warming, cooking, protection from wild animals, etc., were uses

DOI:10.4324/9781003203100-1

that led to the idea of energy sources. One could say that this was the threshold of science.

Conjectures like these tell us of scenarios that led to the development of science and mathematics. The knowledge so acquired led to improvements in the life of the caveman, who must have decided that it was time to move out of caves and have his own abodes. He thus was embarking on the so-called civilized life. In this book, we will start somewhere here. We will discover how the parallel growths of mathematics and science took place and how they helped each other and contributed to man's advancement.

What should one expect from a law of science?

As we noted earlier, the curiosity about how nature functions led to the development of science. Many natural phenomena indicated that nature does not function randomly but has some sort of pattern about it. Indeed, man also learnt the trick of using the pattern to his advantage. For example, growing food grains on his farm required knowing the seasonal change through the year; the 'year' itself was a period over which the view of stars and constellations in the sky underwent one cycle. The incentive, therefore, was to uncover the way nature uses a particular pattern.

We will also note how a study of numbers led to various branches of mathematics, in such a way that one could play not with numbers alone but also with symbols. Although the use of numbers and symbols led to interesting exercises, there were situations where one could find a use for understanding the patterns in nature. In this way, the 'pure' and 'applied' aspects of mathematics began to develop.

Thus, we shall see works of Fermat, Descartes and others which may have started as results of playing with numbers and symbols, but many ended as pure aspects of mathematics. So, referring to a theorem in pure mathematics, one could ask if it had any use. While the normal answer is "No," one should not rule out the possibility of some application of the theorem in the future. We will give specific examples later in the book where very abstract forms of mathematics turned out, unexpectedly, to be of good use.

We will also describe the empirical results that Johannes Kepler derived from a mathematical analysis of Tycho Brahe's observational records; the unanswered question at the end of that work was, Why do the planets move around the Sun in the way they do?

This question was finally answered by Isaac Newton. It would not be an exaggeration to say that Newton was responsible for setting up the *modus operandi* for the scientific study of nature, a modus operandi that is followed even today.

The poet Alexander Pope has paid tribute to Newton in the following verse:

> Nature and nature's laws lay hid in night
> God said, "Let Newton be" and there was light.

It is very rare for a literary personality to pay such high tribute to a scientist. But Newton's work was of such high standard and enjoyed such public visibility that it very well deserved the accolade.

Till then, amongst the persons whose work had proved to be such wide influence was Aristotle. We will see how Aristotle's reasoning prevailed over others, including such stalwarts as Copernicus, Galileo and Kepler. Eventually, most of Aristotle's tenets proved to be wrong. Then what was the correct version? So far as *dynamics*, that is the science of motion, was concerned, it was provided by Newton's three laws of motion.

This then has been the pattern. Human curiosity prompts answers to questions about nature leading to laws of science. The laws are subject to change if new findings contradict their predictions. So we have the changes brought about by scientists from time to time:

Aristotle – Copernicus – Galileo – Newton – Einstein – Quantum theory – ?

The sequence does not end but shall continue as science advances. What about mathematics? It also expanded and added to its knowledge base. The original view of mathematics may have been its applied side. But as one thought deeper, one saw that a much richer picture emerges if one takes both the pure and applied sides together. In this book, we shall continue to emphasize this aspect.

1

GLIMPSES OF GREECE

We begin with Greece, not because it is one of the oldest civilizations known but because we have more details about ancient Greece available. Greece has always been a small country, but its inhabitants have been more adventurous and outward looking. History tells us that Alexander the Great set off from here to conquer the world. And he reached the north-western part of India before his army, tired of long journeys and missing their near and dear ones, prevailed upon him to turn back. But apart from Alexander, there were others known for their scholastic achievements whose influence lasted much longer than the conquests of Alexander. We may mention some here. There were philosophers like Plato, Socrates, Aristotle, etc.; practical scientists like Thales, Eratosthenes and Archimedes; and astronomers like Aristarchus, Ptolemy and a woman scientist like Hypatia.

The scholars described here are only given as examples. Likewise, the episodes mentioned are indicative. More exhaustive accounts are available elsewhere, as described in the references at the end of the book.

How tall are the pyramids?

The story dates back to Thales from Milatus, a small town about 320 km from Athens. Although mathematics as a subject had still to have an independent existence, Thales had already developed a reputation for its application. The following anecdote illustrates the clever way Thales put it to use.

Thales had gone on a tour of Egypt with a group of friends. Naturally, their itinerary included the giant pyramids. The guide wanted to impress these 'foreigners' with the enormous size of these monuments. "The base of this pyramid is estimated at about 518 cubits (a cubit may be taken as approximately 45 cm)," said the guide, and the tourists were suitably impressed.

DOI: 10.4324/9781003203100-2

Except one! The redoubtable Thales had a question: "What, pray is the height of this, admittedly enormous, object?" This was an unexpected query, and the guide did not have a ready answer. He begged leave to check with his superiors who might have the data somewhere in their records. They, however, did not readily have any record of the height of the pyramid. As they were searching, they heard a shout from outside: "I have found the answer, Mr Guide. I estimate it as about 321 cubits. Come here, and I will explain how I did it."

How did he do it? Thales explained by using a technique that in modern times is known as use of similar triangles.

Draw two similar triangles ABC and PQR, where similarity means that the corresponding sides have the same ratio. Thus,

AB/PQ = BC/ QR = CA/RP.

Figure 1.1 illustrates this result. Notice that the two triangles ABC and PQR look 'similar in shape,' but the first one looks bigger. If we blow up the second triangle so that its side PQ equals AB, then we will find that the other sides also match – i.e., QR equals BC and RP equals CA.

Thales used this trick! In broad sunlight, he stood on plain ground so as to cast a shadow whose length he got measured. As shown in Figure 1.1, the triangle PQR has side PQ for Thales's height and side QR for his shadow. Now Thales argued that provided we have the shadow of the pyramid as BC and its height as AB, we must have the aforementioned result for similar triangles

AB/PQ = BC/QR.

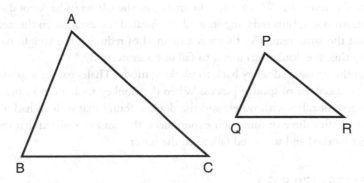

$$\frac{AB}{PQ} = \frac{BC}{QR} = \frac{AC}{PR}$$

FIGURE 1.1 Measuring the height of the pyramids. The tourist taught the guide to measure the shadow and infer the height of the pyramid.

This result holds provided measurements are taken over a short time because the length of the shadow changes with time. Since the pyramid base was well measured, it was not difficult for Thales to locate the shadow of the pyramid top and measure the base side BC.

When even donkeys teach you a lesson

The same Thales learnt a lesson from his observation of the behaviour of a donkey! This happened in his capacity of processing the product of a salt mine he owned.

Thales's mine produced salt in the form of impure 'stones' of salt and soil. These had to be purified so that the final product could be sold as pure salt. Thus to begin with, these lumps of pristine salt had to be collected and sent to the facility where purification was carried out. Trains of donkeys were used to transport impure salt lumps on their back to the purification facility. The route of transport included going through a shallow path across a stream.

On one such occasion, a donkey carrying its load through the stream slipped and fell in the water. It got up and continued its journey as part of the donkey train. However, the event was repeated the next day with the donkey falling in the water. It raised itself and proceeded as before. Not only that, the event repeated itself every day, and this puzzled the supervisor of the donkey train, and he brought the matter to his employer's attention.

Thales, the employer, was also puzzled by the event and decided to observe it happening. He called the supervisor to watch too. And Thales was amused to watch the process in action. The donkey (contrary to the popular view of its intelligence) was a clever one. When it fell the first time, the salt on its back got dissolved in the stream water, thus reducing its load. It repeated the exercise on the next trip and found the same result. So this was a method of reducing the weight to carry. Realizing this, the donkey arranged to fall in water every day!

To get the wayward donkey back to working mode, Thales gave it a special load to carry: collections of sponge pieces. When the donkey took a dip in the water, the sponges got filled with water, and the donkey found that its load had *increased considerably*. After three or four such experiences, the donkey realized that his trick no longer worked and so ceased falling in the water.

Euclid (325–270 B.C.)

The geometry we are taught in school was set up as a branch of mathematics by Euclid. A formal structure starting with *axioms* and developed with a series of theorems characterizes Euclid's geometry.

Euclid was around in the period when the Greek civilization was enjoying intellectual prosperity but whatever details are known about Euclid are because of the records kept by the Arabs. From those records, we are able to determine Euclid's biography. According to this valuable source of information, we know that Euclid was born in a place called Tyre, which today is part of Lebanon. His father was Nocratis, and his grandfather was Zenarchus.

For higher education, Euclid went to the school where Plato taught since it was famous as a place of intellectuals. The admonition at the front entrance to this academy said, "Those who do not know mathematics may please not enter this place." Euclid had no difficulty with this condition, and for a few years, he studied and later taught at this place. Then he moved to Alexandria, which was considered an intellectual centre. He lived here all the rest of his life. Here he wrote his magnum opus the *Elements*, which even today is considered an excellent introduction to geometry.

The book is in 13 volumes, although Euclid calls each chapter a 'volume.' Of these, the first four talk about geometry of the plane, while the next three deal with geometry of three dimensions (called solid geometry today). The rest of the six chapters deal with geometry and arithmetic. Based on his axioms, he proves no less than 465 theorems.

What are axioms? These are statements assumed to be true and the whole facade of theorems is based on these. The axioms appear reasonable, and one can take them as self-evident truths. For example, one axiom says that if two things are similar to a third thing, then they are similar to each other.

This statement appears to be true, although one must admit that its truth cannot be proved by using the rest of the axioms. There was, however, one axiom which led to considerable discussion and arguments as to whether it could be proved with the help of the rest of the axioms. This axiom relates to parallel lines. If we have a straight line *l* and a point *P* not on it, we can draw one and only one straight line through that point parallel to the given line (Figure 1.2). This axiom seems reasonable but can one prove it otherwise?

By calling it an 'axiom,' Euclid assumed that it could not be proved with the help of other axioms. But many other mathematicians thought that it could be proved; although they could not prove it! This axiom is often referred to as *Euclid's parallel postulate*. We will come back to it later in the book because it was destined to play an important role in both mathematics and science.

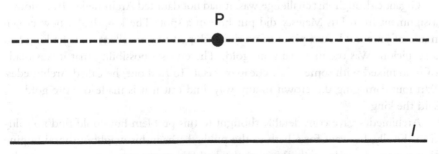

Euclid's Parallel Postulate

FIGURE 1.2   Euclid's parallel postulate

King Ptolemy tried to learn geometry but was put off by Euclid's dry formal style. He asked, "Is there no easier way of learning geometry?" To that query, Euclid is believed to have replied, "Sire, there is no royal road to geometry."

Why eureka?

The name Archimedes is known to most of us because we encountered his name in an old episode in ancient Greece narrated in school textbooks. However, most accounts end halfway with the word "eureka" called out by Archimedes. It is therefore worthwhile to cover the entire episode.

It may be appropriate to begin by giving a brief introduction to Archimedes. His lifespan was between 287 to 212 B.C. He was exceptional as a scientist because his work was of a practical nature, often providing an improved and easier way of carrying out the job at hand. One example will suffice to illustrate this claim.

Archimedes was born in Syracuse. Although he could have continued working in the famous city of Alexandria, where he had gone for higher studies, he was devoted to his mother city and returned home to work there. He had a reputation for providing (science-based) practical solutions. Thus, one day, he received an SOS from his king. King Heron had got a luxury ship constructed, which was the ultimate in having gadgets to make life on the ship most comfortable. The trouble was, however, that having made the huge ship on dry land, how could one shift it to the sea nearby?

The king was getting impatient and felt frustrated that for want of transfer to seawater, his favourite toy would be unusable. That is when he remembered Archimedes and recalling his prowess, summoned him to provide a solution to his problem. The scientist looked at the huge ship but was undaunted. Instead, he asked the king to load the ship with passengers and other service items while he worked at the solution.

He finally made a contraption of pulleys and levers attached to the ship and provided its end point with a rope which he requested the king to pull. How will a pull by one man move the gigantic ship towards water? Somewhat incredulously, the king did as he was requested, and lo and behold, the ship moved as required.

Gigantic though this challenge was, it had not daunted Archimedes. But another assignment from His Majesty did put him in a spot. The king had a new crown made of pure gold. As a piece of workmanship, it was excellent, but the king had a suspicion. Was the material pure gold? There was a possibility that it was made of gold mixed with some other cheaper metal. To find out, he called Archimedes. Without damaging the crown in any way, find out if it is made of pure gold. So said the king.

Archimedes gave considerable thought to this problem but could find no solution. Finally, he went for a bath in the public baths in his neighbourhood hoping that a relaxed mode of thinking in the bathtub might provide a solution to his problem. Sure enough, as he climbed into the water-filled tub, some water spilled out. Archimedes had witnessed this many times in the past, but that day had a special significance for him. And he gleefully shouted out, "Eureka," meaning that he found the answer. And, to settle the answer to his problem, he jumped out and ran naked to his house to the shock of witnesses on the street.

Was the gold in the crown genuine? How did Archimedes find out?

Archimedes shouted out 'eureka' when he found a 'non-invasive' way of testing the crown. Suppose he weighed the crown and found that it was M grammes. By dipping the crown in water, he could find its volume. This was the information he needed and was pleased to find that the overflowing tub could give it. Of course, he did not need a bathtub; a typical bucket would do. By filling the bucket with water to the brim, and then gently inserting the crown, he would collect the water displaced by the crown and measure its volume. Suppose this volume was V.

Now he used the fact that gold was the densest material known. (Today, we do know of material like osmium, which is denser than gold, but this information was not known either to Archimedes or to the crown maker.) So pure gold would have the maximum density then known. Let it be D. So the crown was expected to have density D. In actuality it was $M \div V = K$; to decide the issue was easy. For the material of the crown to be declared pure gold K and D *must be equal*. If it was found that D was greater than K, then it implied that the crown contained gold contaminated by some lighter material like bronze or copper.

This procedure, therefore, justified the exclamation, "Eureka."

Hypatia: a woman scientist

Hypatia from Greece may well have been the first woman scientist ever. She was born in the year 370 A.D. but lost her mother at a young age. Her father, Theon, however, saw to it that the little girl did not suffer for want of anything. Theon encouraged her in various outdoor activities like running, jumping, horse riding, hiking and mountaineering. While he saw to it that Hypatia was good in her studies, he also encouraged her to be good at elocution, general reading, etc.

Theon was himself a teacher of astronomy and mathematics. His place of work was in the famous museum and library of Alexandria. Apart from studying in that famous city, encouraged by her father, Hypatia also travelled in the neighbouring regions to pick valuable experiences.

When she returned from her travels, Hypatia joined her father in the studies of mathematics and philosophy. She also did valuable work in improving textbooks. Those were days before printing was discovered, and so books in handwritten forms were much in demand. New or modified texts in handwritten forms were acquired by visiting scholars since Alexandria was known to be a source of such books.

The ancient Greek scholar Ptolemy had used certain tables in his astronomy writings, and over the years, they needed to be improved and updated. Hypatia undertook that valuable task and completed it very well. For example, chords subtending by an angle as small as the 3,600th part of a degree were evaluated. She also wrote a commentary on Apollonious's classic book on conics.

Her scholastic work spread her reputation as a teacher and orator, and she was much in demand for lectures. Unfortunately, times were changing, and Alexandria was no longer as cultured a city as it used to be. Antisocial elements were on the rise, and Hypatia was dragged into a controversy between religious leader Cyril and the Police Head Ostence. In a street brawl arising in this fight, she was attacked while on her way to give a talk, and she was brutally killed.

As a woman scientist, she was unique.

Homage to Aristarchus

Does Earth move around the Sun or does the Sun move around Earth? In our school texts, we are told that the first alternative is the right one, and it was Nicholas Copernicus who had the courage to say so. Why did making such a statement require courage? Because for several centuries, the Holy Scriptures had made it a religious tenet that Earth is stationary. To say otherwise, such as implied by the previously mentioned first alternative, amounted to sacrilege.

It was even before Christianity that the belief in fixed Earth was prevalent. It was Aristarchus of Samos (310 B.C. to 230 B.C.) who had the courage to go against the fixed Earth hypothesis. He was asked Why? How did he prove that Earth moves? To settle this issue, he suggested a scientific test. If we look at the background of stars in the night sky, we will find that if we are on a moving Earth the relative directions of nearby stars would change. The effect would, of course, diminish with distance. Thus, distant stars would appear stationary even if we were seeing them from a moving Earth. Since the noticeable change of direction will happen for nearby stars, Aristarchus estimated the effective change of direction for some stars, which he believed to be nearby.

Aristarchus, therefore, used the distances of nearby stars as estimated by him and predicted how much change of direction occurs for them. He asked the observers of the sky to check if such a change of direction occurs over a few (ideally six) months (Figure 1.3). They did not find any such change and therefore rejected the 'moving Earth' hypothesis.

Prediction of Aristarchus

Why was Aristarchus wrong in his prediction?

Aristarchus was using the 'parallax' test for his hypothesis. In Figure 1.3, we have two positions for the earth A_1 and A_2 during its orbit of the sun. These are used to view the Star S. The distance between these two positions is two astronomical units. One astronomical unit equals 150 million km. This is the Earth–Sun distance, assuming (as Aristarchus did) that the orbit is circular. So we have the distance A_1A_2 equalling 300 million km. We then estimate the angle A_1SA_2, which tells us the change in the direction of the star when seen from these two vantage points. The typical distance between Earth and some nearby stars is of the order ten

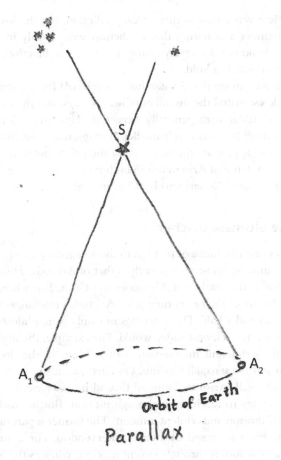

S

A₁   A₂

Orbit of Earth

Parallax

FIGURE 1.3   Aristarchus's test. If Earth moves, then after six months, its location would have changed maximally. So Aristarchus expected to observe some change in the directions to nearby stars.

light years. A light year is the distance travelled by light in one year, and it equals approximately 10 million million km. So the distance from the aforementioned star to us is 100 million million km.

If we take a million kilometres as a unit of length, then our triangle in the figure has the side SA_1 equalling 100 million million and the same for the side SA_2. The third side of the triangle is very small in comparison. It is 300. This makes the vertex angle $A_1 SA_2$ very thin. Using the radian measure, we get it as small as three parts in a million.

Aristarchus did not have data on the stellar distances. His guesswork was a gross underestimate of the distances, and so he got somewhat smaller values for stellar distance. Apart from this, the measuring instruments were not very sensitive, and they ruled out the angle predicted by Aristarchus.

So the problem was sound in theoretical prediction, but the instruments available for measurement and testing that prediction were grossly inadequate. Since the instruments could not detect any change in the stellar direction, the geocentric point of view continued to hold sway.

In hindsight, we can say that Aristarchus was rejected for the wrong reason. He had grossly underestimated the distances of nearby stars, and the observing equipment was much cruder than generally expected. Thus, the effect predicted by Aristarchus was there but was much smaller in magnitude. We will come back to this episode later to appreciate the smallness of the effect involved.

Today, there is a statue of Aristarchus in his hometown with the remark after his name that he anticipated Copernicus by 17 centuries!

Aristotle: the ultimate teacher

Of all the names one can think of that led to the intellectual prosperity of Greece, the name that stands out most prominently is that of Aristotle. History books may describe Aristotle as the teacher of Alexander the Great. But whereas Alexander's empire did not last more than a century or so, Aristotle's teachings lasted for longer than a millennium and a half. The teachings not only were philosophical but also talked about nature as a scientist today would. For example, the notion of motion. There are so many different manifestations of motion. In the sky, there are the Sun, the Moon and the stars, all moving. On Earth, there are motions of different objects great and small. Aristotle believed that all bodies have preferred positions in space, and they try to reach them through motion. But he made a distinction between 'natural' motion and 'violent' motion. The former is part of nature's activity, whereas the latter is caused by human intervention. For example, an arrow shot out from a bow moves through violent motion, whereas the Sun's motion is natural.

He further made the distinction that natural motion is in circular trajectories. Since a straight line may be looked upon as part of a circle of infinite radius, that also is allowed under the category of natural motion. What is so special about circles? Aristotle would argue that a circle is the only trajectory in a plane with the symmetry property – namely, if we cut out any part of a circle, we will have a circular arc that fits exactly on the rest of the original circle. In modern jargon, a physicist would describe the property as having translational symmetry. Of course, there were many such ideas that distinguished Aristotle's teachings.

These ideas were accepted uncritically by succeeding generations of scholars who simply memorized and propagated them further, with the result that they became accepted dogma in European countries beyond Greece. The real challenge to them came from Galileo Galilei from Italy in the seventeenth century. Galileo may well be called the first experimental physicist! For he believed in accepting a statement only by experimental verification. We will encounter Galileo later in this book.

PHOTO: Galileo

Galileo's lectures, as well as a book he wrote, gave examples of this new approach. To followers of Aristotle who excelled in talking, this notion of experimental verification was new and not acceptable. However, gradually, his point of view and method of reasoning became current, and Aristotle's ideas were phased out.

The destruction of the Library of Alexandria

The Alexandria Library had a worldwide reputation as a source of literary information. Unfortunately, no details are available as to when and why the library was destroyed. There are several scenarios but one that is often quoted is the destruction by the army of Julius Caesar. It is said that as he was pursuing Pompey into Egypt, he passed through Alexandria. There he saw an Egyptian fleet and ordered that it be burnt. The resulting fire spread and consumed buildings in the city, including the famous library. Thus, several important manuscripts were irretrievably lost.

2

THE EASTERN INPUTS

The question naturally arises as to what was the status of knowledge in the East: in the Indian subcontinent and China. A brief description of the former is given in this chapter, but the reader is referred to the book The Scientific Edge *by this author for a longer description.*

Shulva Sutra

The description of early days invariably begins with the *Shulva Sutra* belonging to the Vedic times (c.1500 B.C.–200 B.C.). The title itself means 'rules of measurement' since the lengths were measured by ropes, the word *shulva* later came to mean 'rope.' The *Shulva Sutra* contains arithmetical/geometrical statements like Pythagoras's theorem. However, the proof of Pythagoras's theorem is not given. Indeed this was a difference in the approach of the Greeks (who believed in giving proofs) and the Indians who were interested in the outcome and not whether it had a proof.

Good approximations for irrational numbers like π and $\sqrt{2}$ were given but no background details of how these approximations are obtained. Thus, the square root of 2 is approximated by

$$1 + 1/3 + 1/3.4 - 1/3.4.34 \ldots$$

a good approximation! But no clue is given as to *how* this result was obtained.

The Shulva Sutra belongs to *VedangKalpa*, the sixth part of the six Vedic divisions. The fifth part is known as *VedangJyotisha*, and it concentrates on astronomy.

The Vedic era was best known for the knowledge of zero and the decimal system, which was more convenient than the Greek way of writing numbers. In this

DOI: 10.4324/9781003203100-3

connection, the late Professor Abdus Salam made the following statement at an International Conference held in Dhaka in 1967:

> *Abdullah Al Mansur, the second Abbasid Caliph, celebrated the founding of his new capital Baghdad by inaugurating an international scientific conference. . . . The theme of the conference was Observational Astronomy. . . . He wanted and he ordered at the conference, a better determination of the circumference of the Earth. No one realized it then but there was read at the conference a paper destined to change the whole course of mathematical thinking. This was a paper read by the Hindu astronomer Kankah, on Hindu numerals, then unknown to anyone outside India.*

The unfortunate aspect of this is that there are very few written records of Indian contributions, and one has to depend on what Arabs had translated for their studies and the documents available with the Chinese. We shall encounter an example of the latter type shortly. Of the few records available may be mentioned the *Bakshali* manuscript. This 70-page manuscript was unearthed in 1881 in the famous archaeological site of Takshashila near the city of Peshawar in Pakistan. The 'pages' are *bhoorjapatras* (birch bark) and the script is called *Sharada*, the language being a dialect of *Prakrit*, which itself started as a dialect of *Sanskrit*. Archaeologists estimate the date of the manuscript as around 200 B.C. or so. The mathematical operations described in the manuscript include quadratic equations, arithmetical and geometric progression, square roots of general numbers, etc., thus indicating a high level of the subject prevalent at the time.

We will now see how astronomy can help in dating. The Pole Star plays a key role in the calendar provided by astronomy. The Pole Star lies very close to the spin axis of the spinning Earth with the result that while other stars go round the axis, the Pole Star keeps to its place.

An immovable location?

There is an Indian legend inspired by these astronomical observations. The observations are about the starry sky in which stars move from east to west. The rising and setting of stars all share this property. However, there is apparently one star that does not move. Its location in the sky can be specified precisely, but the important result that emerges is that the answer does not change from night to night even over several days or months.

The star in question is called *Polaris* or the Pole Star. But its name as given by the Indian observers is more revealing. Called *Dhruva*, it means something that is fixed. This name also comes with an old legend as follows.

There was a king called *Uttanapad*, who had two queens: *Suniti* the senior one and *Suruchi* who was the junior but favourite one. The latter threw her weight about knowing that her husband would always support her.

One day, the king was playing with his two sons, Dhruva the son of Suniti and Uttam, the son of Suruchi. He was carrying Dhruva on his lap when Suruchi turned up and told Dhruva to get down and make way for his younger co-brother. The order and the insulting way it was given made Dhruva very sad, as well as angry. He quit his princely abode and went to the forest for a long and hard penance. He continued in this hard exercise until Lord Vishnu (the second of the Trinity of Gods) was pleased to offer Dhruva what he wished for. "Please put me in a place in the heavens from where no one can shift me." Vishnu was pleased to grant him this wish, and that is how he is in the sky fixed in place.

Although the legend does not talk of the slow movement of the Pole Star, the ancient Indian astronomers were aware that Dhruva's location in the sky was slowly but surely changing. We will return to this aspect shortly.

Even Dhruva moves!

The legend of Dhruva arose out of the observation that the Pole Star does not move from its observed position. While this may be so for night-to-night observations, very long-term observations show that the star does change its position. What does this mean?

Spinning top and spinning Earth

The axis of the top generally processes. So the direction of the axis changes its direction in space. In Figure 2.1(b), a similar precession causes the Pole Star to lose its prime status: the other stars change their position in the sky, but now the Pole Star, which stayed fixed, is subjected to a small change like the other stars. The figure on the side demonstrates this.

A simple experiment will explain the observed phenomenon. Take a top which can be set moving by winding it with string and pulling the string swiftly. The string pulling will generate spin in the top, and if it is set to move on the floor, it will in general move with its spin axis at an angle with the vertical direction as shown in the adjoining figure. In fact, the spin axis of the top describes a cone. The spin of Earth is somewhat similar. Instead of spinning on a hard floor, it spins in space as it goes around the Sun.

If we imagine ourselves on such a spinning Earth, at any time, the spin axis will closely point to the Pole Star. This gives us the impression of the Pole Star remaining fixed in space. But like the toy top has its spin axis going round describing a cone, so too does Earth's spin axis. It moves a lot more slowly than the axis of the spinning top. The cone mentioned above is described in about 26000 years.

As a result, the position of the star Polaris, which is currently enjoying the status of the Pole Star, is slowly changing, and over the course of time, another star may better qualify for that position. The faint star Thuban occupied the Pole Star status around the year 3000 B.C. Our own Pole Star will be closest to the point where Earth's spin axis intersects the sky in the year 2102 A.D. Later in the year around 4200 A.D., the star Errai will enjoy the Pole Star status while later still around 7500 A.D., the honour will go to the star Alpha Cephei.

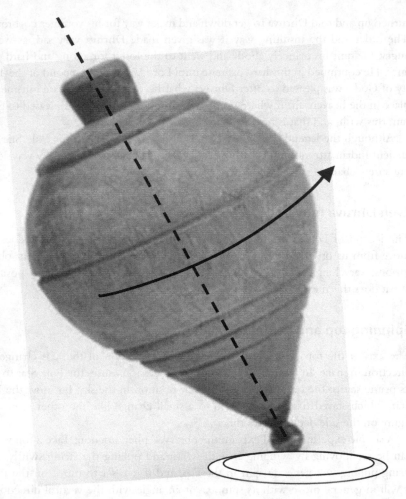

FIGURE 2.1(A) Spinning top has an axis of spin. For spinning Earth, the axis points to the Pole Star.

So Dhruva (of the legend) might feel let down after all that long penance!

A celestial calendar

This slow turning of Earth's spin axis gives us a long-term calendar. The Indian nationalist and scholar Bal Gangadhar Tilak used this calendar in dating Vedas. The principle behind his approach was as follows. In the Hindu holy book *The Bhagavadgita*, one chapter describes the select or the best of any group of objects animate or inanimate. In that context, it is stated that amongst months of the year, *Margasheersha* (an autumn month) is the best and amongst the seasons the

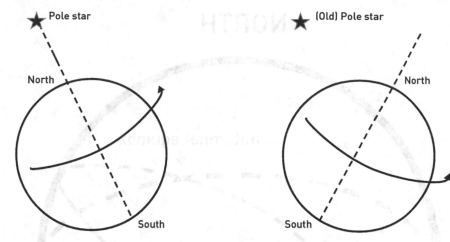

FIGURE 2.1(B)   The Earth's axis of spin processes. Then it does not point to the Pole Star.

best is spring. But this stipulation implies that the best month does not fall in the best season! To make the statement logically consistent, Tilak argued as follows. Because of the slow precession of Earth's spin axis, the calendar months change progressively. So he calculated how long ago the month of Margashirsha came in spring.

He got an answer prior to 4000 B.C., much different and higher than from other methods of determining age. The method used by him was correct, but the literary references he used were challenged. Anyway, the method as such is useful in dating astronomical references.

A projected view of the cosmos

It is useful to look at stellar motions as projected in the (distant) sky as seen by the typical observer. In fact, the stars at night that we see are such projections on the sky. Our own position identifies a celestial equator (Figure 2.1(c)).

Figure 2.1(c) shows two circles intersecting. One is the circle traced by the Sun as seen from Earth throughout the year while the other is the equator as projected on the sky. Thus any astronomical observer has two fundamental circles projected on the sky serving as references. The two circles intersect at two points γ and Ω and are called the vernal and autumnal equinoxes as shown. Because of Earth's spin precessing as shown earlier, the two points slowly move on the two circles. This in turn leads to a change of seasons on a long-term basis. As the Sun moves around the Earth throughout the year, the different parts of the full circle can be identified by groups of stars there, with there being generally 12 such divisions. These are known as zodiacal signs.

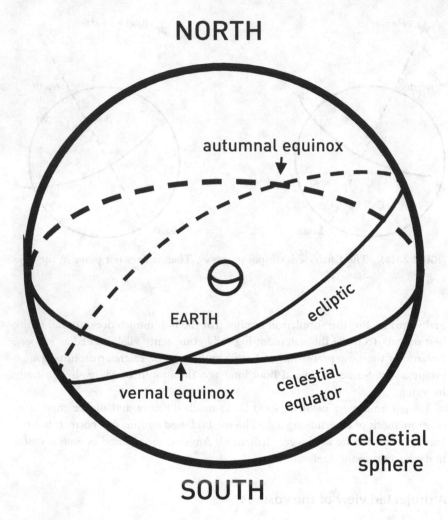

FIGURE 2.1(C) The celestial equator and the ecliptic projected on the celestial sphere

That is how, for example, the Makar Sankranti Festival slowly comes later in the year. It is supposed to occur when the Sun enters the 'Makar' constellation (the sign of Capricorn).

The golden age of Indian science

As mentioned earlier, compared to the Greeks, the Chinese and the Arabs, the Indians did not keep systematic records of ideas or events. The educational tradition in India was mainly oral: each generation of scholars orally repeated what their teachers taught as the original knowledge from the Vedas. Thus we have very scanty knowledge of what was taught in the early times. From what is known, we have a tradition of scholarship dating from Aryabhata I to Bhaskara II, dating approximately from the fifth century A.D. to the twelfth century A.D. This was a period when Europe was in the dark ages and could not claim scholarship of a high order. The period, on the other hand, shows that there was scholastic leadership in India for which reason the above period may be called the *golden age* of Indian science and mathematics.

As we mentioned before, the biodata of even the leading scholars in this period are very limited. We will therefore have to be brief, except possibly for the last scholar Bhaskara II for whom data exists in the books written by him.

Aryabhata I gives his birth year as 476 A.D. in his book the *Aryabhatiya*. The book itself was written in the year 499 A.D. The book is mainly about spherical trigonometry and astronomy. Amongst its specialties is the sine table (called *Jya* in Sanskrit) for angles at intervals of 3^0, 45 minutes of arc. In one of the verses of *Aryabhatiya*, the author says that just as someone going on a boat sees the trees on land going in the opposite direction, so does Earth's spin cause us to view the stationary stars as moving westwards. This claim was against the prevailing belief (originating with Aristotle) that Earth is stationary and the whole canopy of stars revolves around the polar axis. Although there is no detailed discussion of how Aryabhata's idea was received, what little is known shows that the idea may not have been well received.

There is reason to believe that Aryabhata wrote another volume on astronomy later in his life. But, unfortunately, the manuscript of the book did not survive. We will skip across the later scholars like Varahamihira, Brahmagupta, etc., during the golden age and highlight some interesting facts about the last in the series: Bhaskara II.

Bhaskara II: astronomer and mathematician *par excellence*

The year 2014 marks 900 years since the birth of Bhaskaracharya II, a renowned astronomer and mathematician. (The title 'acharya' added to the name indicates that the person was a well-known teacher.) Bhaskara was the last in a series of Indian savants spanning the golden age, starting with Aryabhata in the fifth century. This period of seven centuries marks the golden age of Indian science, a period when Europe was in the dark ages, China isolated from the international traffic of knowledge and Arabia played the role of translating and transmitting that knowledge. Indeed, the Arabs played a very useful role in this capacity, and some of the Greek manuscripts which were destroyed in the sacking of the Alexandria Library survived in their translated version. Although the Kerala school of mathematicians continued its very valuable work in later centuries, the continuity of the tradition of astronomy coupled with mathematics lasted mainly until Bhaskara II.

As there were celebrations for the ninth centenary of Bhaskara, it is worth recalling some of his achievements and viewing them from a modern perspective. A beginning was made by a couple of meetings arranged by the Marathi Vidnyan Parishad, a non-governmental organization for science popularization in the

Marathi language, headquartered in Mumbai. The first meeting was held on January 19 in the small town of Patan near Chalisgaon, a town in Maharashtra on the Mumbai Kolkata train route. The second meeting was in Jalgaon, a bigger city near Chalisgaon. In both places, the participation of student audiences was arranged so that the new generation was made aware of this bright phase in India's distant past.

We take this opportunity to highlight some of the exceptional features of Bhaskara's work. He had four separate volumes in one collection called *Siddhanta-shiromani*, which included books on arithmetic, algebra and geometry, as well as astronomy. Perhaps the best known of his works is *Lilavati*, which is a collection of mathematical problems addressed to a lady of that name. There is no unanimity as to who she was. A more common version is that she was Bhaskara's daughter who became a widow at an early age and spent her life working on mathematics with her father. Expressions like 'Dear Child Leelavati' in the text seem to corroborate the daughter version, but there has been no last word in the controversy. But the enjoyment of the book should not be spoilt by worrying about who Lilavati was. Those who are accustomed to the (erroneous) belief that maths is dry and tasteless will change their mind after reading the highly interesting problems written in the excellent poetic genre.

A poetic problem

Consider this one, for example: "In the rainy season, ten times the square root of the number of a group of swans together with one eighth of the total number flew from the water reservoir to Kamalini forest nearby, leaving three pairs sporting together in the water decorated by lotus flowers. Dear Child, tell me how many swans are there in all." In case the reader wants to check his or her solution, the answer to this problem is 144.

Though there is no historical backing for the following legend, it is nevertheless worth narrating.

As was usual in those days, astronomers also carried their calculations to *astrological predictions*. It is said that Leelavati's birth chart indicated that her marriage would be disastrous to her bridegroom, leading to his untimely death. Bhaskara's 'astrological calculations' however indicated that there was an exception to this prediction provided the wedding took place at a precise hour, which Bhaskara had calculated.

Not to miss that 'one and only' chance for a successful marriage for Leelavati, he set up a very accurate water clock and asked Leelavati to watch it and give him advanced warning as the auspicious hour approached. While Leelavati waited and watched, the necklace she was wearing broke, and the beads in it spread around. One bead went as far as the slot in the water clock and stopped the steady flow of water in it. Unfortunately, Leelavati, who was engrossed in thinking presumably about the imminent change in her life, did not notice it. So the auspicious moment came and went, and when Bhaskara came to check, he noticed that it was too late to carry on with the wedding ceremony.

Poetic though they are, Bhaskara's problems conceal a sophisticated level of algebra and geometry. Indeed, posing problems and solving them seems to be a favourite pastime of Bhaskara, suggesting that he was a good teacher, as well as a research worker. The aesthetic and poetic aspects of his writings could perhaps be traced to his place of origin, for the locale of Bhaskara's birthplace is also picturesque, being overlooked by hills and forest. This location may have played a role in firing the literary talents of the young Bhaskara. Thanks to the biographical notes made by his grandson about the family, some interesting details are known about Bhaskara. For example, his father was also an astronomer. In fact, several of his relations were scholars in different fields, perhaps like the mathematically endowed Bernoulli family in Europe some six centuries later. (See Chapter 7.) Bhaskara was a classic scholar and interested in many other intellectual pursuits.

To appreciate the level of Bhaskara's work, we need to go to the seventeenth century – that is, around six centuries *after* Bhaskara. That was the period when mathematics was a growing subject in Europe, and its problems attracted many intellectuals. One way the growth was proceeding was by mathematicians asking difficult questions to one another. Mathematicians like Pierre de Fermat, Leonhard Euler and Joseph-Louis Lagrange are credited with many contributions that advanced the subject in this way.

In the year 1657–1658, a particularly difficult challenge posed by Fermat received a reply from another mathematician, Viscount Brouncker, who also happened to be the first president of the prestigious Royal Society in London. A complete and rigorous solution was, however, produced much later by the French mathematician Joseph-Louis Lagrange. Lagrange's solution was very difficult and required a lot of number crunching.

Subsequently, however, it was realized that *Bhaskaracharya had solved this very problem with a very elegant method!* To appreciate the difficulty of the problem, here is a simpler version of it: consider the relation

$$x^2 = Ny^2 + 1.$$

Suppose we are given $N = 7$ and asked to find the smallest pair (x, y) such that the equation is satisfied. A simple 'trial and error' approach gives us the answer, which is $x = 8$ and $y = 3$.

But for some other values of N, the answer is not so simple. Indeed, mathematicians like Pierre Fermat posed deceptively simple problems.

The challenge was to find the smallest pair of numbers x and y which satisfy the previous equation for a specified N. For $N = 61$, the answer turns out to be particularly difficult. The solution is

$$x = 1766319049 \text{ and } y = 226153980.$$

It is clear that a solution like this is not to be found by trial and error. How to arrive at the solution by a systematic method? It turned out that Bhaskara had the answer!

Bhaskara's *chakravala* method leads to this solution in a relatively short and simple way. As the name implies in Sanskrit, this method is cyclic in nature. It will take us too far along technical details to follow it here. We end with the comment by the noted historian of mathematics, C.-O. Selenius:

> No European performances at the time of Bhaskara or much later, exceeded its marvellous height of mathematical complexity.

To what extent did Bhaskara anticipate the law of gravitation? There are statements in his book in which he wonders about a stone falling on Earth: could it have been because of attraction by Earth? He ridicules the belief prevalent those days because of Buddhist scholars that the stone and Earth are both moving through space, and the stone falls on Earth because of its higher speed. He also did not believe the puranik view that Earth rests on four elephants. What do the elephants rest on? If the traditional answer were 'tortoise,' he would ask what did the tortoise rest on. In short, he was mature enough to imagine that Earth need not rest on anything nor move at speed in space. He was comfortable with the idea that Earth floated in space (as it would under no forces). Although the story of Newton and the apple is similar, scientists would want a more quantitative statement or derivation of a fundamental law than a falling apple or a falling stone. History has it that Newton had access to Kepler's data and his derivation of the laws of planetary motion therefrom, and he derived the inverse square law by using his newly created branch of mathematics called calculus. For this reason, and rightly so, Newton is given credit as the genesis of the law of gravitation.

Finally, Bhaskaracharya was not just confined to theory: he describes several *yantras* – that is, instruments of observation. His *Goladhyaya* has a long chapter containing at least ten yantras used for astronomical observations. Some of them have sophisticated arrangements for measuring time. He had indeed appreciated the need for measurements and the importance of time measurements in particular. In the super-specialist world of today, Bhaskara's all-around scholarship stands out.

Astronomy and mathematics from Kerala

It is usually assumed that the mathematical and astronomical activity that peaked around Bhaskara II faded thereafter. This assumption was contradicted by some findings of two officers of the East India Company. As history tells us, the British 'sneaked' into India through the commercial loophole. Indeed, the early incursions into India of this company were purely commercial and did not bring any potential danger to the natives. Gradually, however, the inner motive of colonialism began to emerge.

Benjamin Hain and Charles Whish, however, carried academic and cultural credentials as they landed in southern Kerala. They were trained in mathematics and were interested in knowing what was the level of understanding achieved by the natives as compared to what was understood in Europe.

This was when they noticed that some Keralites were using the notion of infinite series in evaluating mathematical expressions and used them to calculate trigonometric quantities. In Europe, such series were used thanks to Newton, Gregory, Leibnitz and their mathematical followers in the seventeenth century. It was clear that the Keralites had independent access to this kind of mathematics. For example, the series known after Gregory giving expressions for the mathematical quantity was used by them. How did they get these results?

Hain's results were published by John Warren from the Madras Observatory in the periodical *Kalasankalita*. Likewise, Whish gave a presentation on it in 1832 to the Madras Literary Society. This was subsequently published in the periodical of the Royal Asiatic Society. This article contained a lot of information on the mathematical achievements of the Kerala school. Still even today a lot needs to be discovered!

One of the fundamental formulae is the infinite series describing π:

$$\frac{\pi}{4} = 1 - \frac{1}{3} + \frac{1}{5} - \frac{1}{7} + ...$$

This formula was an indicator of the level of sophistication achieved by the Kerala astronomers.

One can therefore see that the classical mathematical tradition in India continued after Bhaskara at least for two to three centuries.

Before we leave the early oriental contributions to science and mathematics, we need to look at the early days when teaching and research were going on institutionally. Thus apart from teaching from scholars like Aryabhata and Bhaskara, what was the facility available for students to receive classroom teaching? In Greece, there were academies where teachers like Plato and Aristotle used to teach. Were there similar facilities in the East, say in the Indian subcontinent? We will now seek an answer to this question.

3
OLD SEATS OF LEARNING

In general, it can be argued that depending on the educational level of human society, it provides for facilities to make education available for its members. We can see the evolution in India, for example. Going back to the Vedic era, *Rishis* – that is, learned wise men – were the sources of education. Typically, a handful of students would live as family members working for the maintenance of the residence (*Ashram*) of the teacher. They would attend teaching sessions at the teacher's home. In the end, while departing from the Ashram after completion of his education, the pupil would offer 'Gurudakshina' to the Guru – it would be some form of money, depending on the paying capacity of the pupil. Thus a low-income student may be asked to pay some nominal amount of money, whereas someone from a princely family may pay in ornaments or pieces of land.

This method of education was appropriately named *Gurukul* – that is, the family of the teacher. Of necessity, these establishments found themselves near each other and also on the banks of a river (as the source of water). Although they were independent, they were eventually merged into larger units which became known as universities. We will take a look at two very old universities in the East followed by one in the West.

Takshashila

The first and oldest university thus came up at Takshashila, which is not far from the town of Rawalpindi in the north-west corner of the Indian subcontinent. I was able to visit the site of the ancient university and see the relics displayed in a museum on the site. The site is in the fertile valley of rivers Jhelum and Sindhu. It was a major town in the state of Gandhar. The town itself is mentioned in the Indian epic of *Ramayana*. It appears that the city was set up by King Bharata to be named *Taksha* after him.

DOI: 10.4324/9781003203100-4

Records show that the city was flourishing well, and a university named Tak-shashila was well-known by the year 800 B.C. When Alexander invaded the Pun-jab, he had already heard of this university, and he took many scholars from there to Greece. Takshashila had a lot of academic activities going on, but its character was not quite that of a university. Each scholar was independent and an institu-tion onto himself. Famous faculty members here included Chanakya, Dhaumya Muni, Nagarjuna and Atreya. Chanakya's book on economics (Known in India as *Arthashastra*) is indeed a classic. His student Chandragupta became the founder of the Maurya Dynasty.

The university had no restriction as to who was entitled to higher education. (In the Vedic times, only high-caste Hindus were entitled to learning.) In Takshashila, the high- and low-level students sat together for their studies. The difference came with how they lived. The poorest lived with their teachers, while the richer ones found houses for greater luxury. The university character of Takshashila came from the wide range of subjects taught there. These included arts, literature, music, phi-losophy, religions like Hinduism and Buddhism, law, chemistry, biology, medicine, astronomy, architecture, accounting and astrology. There were courses on witch-craft and sorcery, as well as on handling snakes.

Being in the north-west corner of India, Takshashila had to bear invasions from foreign attackers like Persians, Greeks, Parthians, Shakas, Kushanas. It somehow managed to recover although deeply bruised. The fatal blow came with the Huns in A.D. 450. The attacks were for possible stores of gold and other valuables and to destroy manuscripts of wisdom which were seen as a threat to the ideas of the attackers!

Nalanda

Although more universities had come up in India after Takshashila, one is restricted in describing them for lack of suitable old records. Kashi (modern names: Varanasi, Banaras) in the north and Kanchi at Kanjivaram in the south played lead roles whenever authoritative certifications were required. Thus disputes in other places of learning were referred to these places for decision. But we have another source of information about Indian universities from the meticulous records kept by a Chinese visitor HuenT'Sang (also referred to as Xuanzang) who came to India in the seventh century B.C. His visit had been motivated by the interaction between the Buddist schools in China and India. But he was especially interested in the University of Nalanda located in today's state of Bihar in the north-east part of India. No account of the Indian academic environment can be complete without some glimpses provided by HuenT'Sang's writings on Nalanda.

He has described the university city of Nalanda as a confluence of Hindu, Bud-dist and Jain religions. Chanakya, described earlier at Takshashila, was born here. Leading religious experts like Nagarjuna, Buddhaghosha, Aryadeva and Jyotipala lectured to a long line of eager students at Nalanda.

In an attempt to revive the old traditions there has recently been established a new University of Nalanda at its old place Rajgir in Bihar. Today, the ruins of Nalanda hardly give any idea of what had existed there. Fortunately, we have HuenT'Sang's detailed records to get a feel of the old days. For example, here Buddha triumphed in his religious disputations with two leading scholars Upali Grihapati and Deegh Tapasi. These scholars opted for Buddhism after losing their arguments with the Buddha. Vardhamana Mahavira spent 14 years of his life propagating the Jain religion here.

Of course, as such institutions grew with scholars of distinction visiting or working here, they also needed royal support for creating various facilities. Thus Emperor Asoka built a Vihara (monastery) to commemorate the birth of Buddha's favourite disciple, Sariputta. The eventual destruction of Takshashila in A.D. 450 left a void which Nalanda amply fulfilled. The Chinese records describe in glowing terms the intellectual and physical well-being at Nalanda. The campus had a very pleasing appearance with palatial buildings, gardens, streams and ponds, boating facility, etc.

HuenT'sang mentions monks living on the campus with a residential four-floor hostel building, with all tall buildings having sky observing facilities. The tall buildings on a misty day appeared to disappear in the mist. The campus had a thick protecting wall around it. The Gupta kings and later King Harshavardhana patronized Nalanda through cash and land donations.

Between the ninth and twelfth centuries, the Pala kings tried to help as much as possible since they had security and other expenses to bear.

Unlike Takshashila, Nalanda was constituted more like a modern university. It had a management council and an academic council with the former represented by a Vice-Chancellor (*Kulapati*) for day-to-day administration. The latter had several scholar members and the committee looked after academic issues in both Nalanda and its sister university Vikrama-Shila situated about 30 km away.

The six Viharas (residential premises) at Nalanda were managed by a committee headed by a Viharpal or a Viharswami who was next to the Kulapati in authority. Each Vihara had its own stamp to deal with legal matters. HuenT'Sang mentions that in his time Nalanda had about 10,000 students and 1,500 faculty. Thus the teacher-to-student ratio was 1:7, a good value from modern standards! Women students were allowed, but there were strict restrictions on where men and women could meet.

Dharmaganja or the Nalanda library was housed in three buildings called Ratnodadhi, Ratnasagar and Ratnaranjak. The first was nine-storeys high, while the other two were six storeys each. The library also published books and preserved rare manuscripts.

The university taught religious, as well as secular, subjects. The students had no fees to pay or any living expenses. They were admitted after tests at the entrance conducted by the Dwarapanditas (wise men at the door). These tests were tough and only 20–30 percent passed. Thus, they were on scholarships once they passed. Nalanda had the top rank amongst educational institutions because it attracted good teachers. Leading names were Aryadeva, Kamalasheela, Karnapati, Chandra-pala, Dantabhadra, etc., and many others.

Tragedy hit this flourishing place through the attacks of Bakhtyar Khilji in the thirteenth century. As in the case of Takshashila, the whole campus was ransacked for valuables, and manuscripts seen as a threat to the attackers' beliefs were destroyed.

The creation of Oxbridge

Just as the vacuum created by the destruction of Takshashila was followed by the prosperity of Nalanda, it so happened that in the thirteenth century, Europe began to feel the urge for intellectual upliftment. And as a consequence, there began to appear on the educational scene new universities. The universities which can be mentioned were Sorbonne and College de France in Paris, Oxford and Cambridge in England and a few others in other European countries. It is interesting that, as we shall see, in the aforementioned pairs one arose because of some defect or problem of the other. Thus, when Sorbonne refused to teach certain subjects, the king of France created College de France with the motto, "We teach everything."

The creation of a university at Cambridge in the year 1209 was likewise because of a controversy at the university created at Oxford a few years before. The controversy had a deep religious connotation. There were several issues of a socio-religious nature that had to be settled. For example, whose privilege was it to make the appointment of the archbishop of Canterbury, the foremost leader of religion in Britain? The king and his followers maintained that the king was the appointing authority. This was violently opposed by the supporters of the pope, the leader of the Roman Catholic religion, with the claim that the pope alone had that power.

Verbal controversies in England on this topic grew and turned violent. Worse still, the arguments led to the killing of intellectuals. Oxford was at the receiving end of such disturbances until those who supported the pope's side left the city for a quieter part of the country. This happened to be the city of Cambridge, a sleepy little place that promised to provide a suitable location for a university. The tiny river flowing by the town was called *Granta* or *Cam*. In due course, a new university came up there and grew. The political atmosphere in the country also quietened over the years and shortly thereafter the pope permitted the graduates of the university to teach anywhere.

It would take considerable space to outline how Oxbridge ('Oxbridge' is a concocted word from the names of Oxford and Cambridge, another less popular such word being 'Camford') grew and prospered. We close this account with a description of the clever two-dimensional way in which the teaching is carried out from early days to even today.

Over several centuries, these universities have grown, and the growth has mainly come through the creation of colleges. It is necessary to clarify that Oxbridge College does not carry out lecturing *en masse*. Rather, the typical college takes care of mainly the accommodation of the pupil. The teaching is carried out by faculties of various subjects. Thus all the teachers of mathematics will be part of the mathematics faculty, and the faculty will arrange lecture courses for students. What do the colleges do? Besides arranging lodging and boarding of the students, they look after their welfare, as well as provide supervisory teaching. Thus besides attending lectures in the faculty, the student spends an hour each week with a mentor as a supplement. This supplement can be very useful in catching up with the coursework. As a general practice, an academic member in Oxbridge belongs to his subject faculty and to a college where he is a fellow.

This two-dimensional system provides a broadening of the pupil's, as well as the teacher's, intellectual horizon. At the college dinner, the fellows eat at the high table, and usually, academics from different fields sit next to each other. Likewise, students from different subjects meet and interact with one another. This system has certainly helped to provide each student a variety of friends from a variety of subjects.

The Mathematical Tripos is the name given to the series of graduate-level examinations in mathematics at Cambridge, United Kingdom. Perhaps with the longest history, these examinations are also famous as very difficult tests of mathematical ability. Traditionally, those who come up to the first-class level are called 'wranglers,' a term presumably reflecting the maturity attained by the student so that he could successfully argue ('wrangle') a point in mathematics. In the old days, the wranglers were ranked with the top one being called the 'Senior Wrangler.' Ranking was abolished in 1909, and today, wranglers are listed alphabetically. Likewise, those in the second class are called 'Senior Optime' and those in the third are called 'Junior Optime.'

The Mathematical Tripos dates back to the early eighteenth century, and mathematics can claim to be the oldest of all subjectwise examinations in Cambridge. Even today, it has retained some of that early aura and traditions. For example, the aforementioned classification is read out on the results day by the chairman

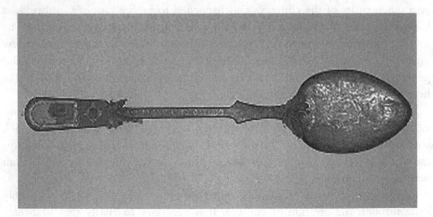

FIGURE 3.1 The wooden spoon presented to the Mathematical Tripos student who got the lowest passing marks. This informal custom was abolished in 1909.

of the panel of examiners who stands on the balcony of the Senate House. This result-reading is carried out at the stroke of nine in the morning from the university church. After reading the list, the examiner throws copies of the list from the balcony. Students who happen to be present for the occasion scramble for them, and some may like to retain a copy for a memento.

Just as the Senior Wrangler used to signify the top student in the batch, the last one in order of merit was called (unofficially!) the 'Wooden Spoon.' When at the convocation, this particular student knelt before the vice-chancellor, and his college buddies used to lower a large wooden spoon between him and the vice-chancellor. In retrospect, one may imagine that being the last on the list but avoiding failure in the examination is no mean feat. It is a pity that this light-hearted event also came to an end when the ranking was abolished.

4

FEEDBACK FROM HISTORY

A pertinent question

The Nobel Laureate Abdus Salam raised the following question. The Taj Mahal in Agra, India, and the St Paul's Cathedral in London, United Kingdom, were two major examples of contemporary architecture completed around the same time. They indicate the high level of architecture achieved in India and Europe around the same time. Yet, history tells us how in the century to come there was a renaissance of science and technology in Europe but no such development occurred in India or the subcontinent. Specifically, in the United Kingdom, there was Isaac Newton whose genius in mathematics and science led to a revolution of ideas and achievements. On the Indian subcontinent, there was no such intervention by a genius. Salam's argument makes us think and ask ourselves the question, "Why?"

This one-word question makes us think and come up with a series of reasons, all of which had contributed to it. We begin with the responses of emperors, kings and no less the large and small maharajas who were known to patronize excellence. Thus there were musicians, artists, poets, litterateurs, etc., who were supported by the local rulers in India. *However, there were no scientists in that list!* The corresponding situation in Europe was quite different.

In England, there was the Royal Society, in France the French Academy and similar societies in other European states. Their origin was in a sense a recognition that science is a force that can be turned to one's advantage. The role of Newton referred to earlier was to demonstrate that there are natural forces that explain the apparently strange behaviour of nature. And following this came the realization that these forces can be turned to one's use. This feeling grew with time and later the so-called Industrial Revolution began to provide practical examples.

DOI: 10.4324/9781003203100-5

In India, unfortunately, there was no one to appreciate this fact. The rulers in India certainly saw that the European tradesmen brought very useful products: but they do not seem to have thought of investing in the science and technology for making these devices. And as a result, the technology market was monopolized by the European traders.

This may be oversimplified reasoning, but it does carry the germ of truth. There are also other reasons which we should not forget. The growth of science led to faster communication, and in the European nations, the colonial attitude spread. By travel over long distances, we can take advantage of natural products; by the power of superior weapons, we can win over tracks of land belonging to others; and by experimentation, we can improve on our medicines. In India, by contrast, the upper-class (Brahmins) Indians were forbidden from leaving the motherland and thus took very little part in explorations of Earth.

Another reason sometimes given for Indian backwardness was religion. Religion can determine a mindset that could be glorified by the concept of plain living and high thinking. If it is suggested that one should limit one's needs to as few requirements as possible, and one may be promised luxuries in the next birth, the incentive to make new inventions to make living easier and more comfortable weakens or disappears. This could have been a reason for going easy on any inventions that make life easier.

There may be other reasons for our "why," but we will end this discussion by bringing in the role played by weather. The weather in India is, by and large, benign, unlike the European weather that can have harsh winters. Fighting with harsh weather, the Europeans were able to advance technology in various ways. With no such incentive in India, that particular route was also not followed.

Another aspect of this issue is seen in the following true story.

The urge to observe

Although the main reason for the interest of the European powers in the development of the Indian subcontinent had been in the land and commerce, there were some Europeans who welcomed contacts with the East for academic purposes. And amongst them were astronomers who were keen to observe the sky from the Indian latitudes.

One such astronomer from France was Guillaume Le Gentil, a reputed observer. It was known that on the dates 6 June 1761 and 3 June 1769 there would be transits of Venus. In a transit, the inner planet Venus or Mercury can be viewed against the background of the solar disc. Like an eclipse, the transit can give important information about the solar system. For this reason, the French government on the advice of the French Academy deputed Le Gentil to India for observing the transit of Venus in 1761.

Although Le Gentil started for India well in advance, he had not allowed for the disruption caused by the Seven Year War between England and France. This meant he had to change route often, and the delay so caused made him arrive in India in the French protectorate of Pondicherry too late for the first transit.

However, Le Gentil did not give up! Rather, he decided to wait for the second event and to use the eight-year intervening period to make important observations.

In due course, when 1769 arrived, disappointment was in store for the French observer. A cloudy sky made it impossible to observe the transit. A somewhat consoling aspect was that the 'rival' British team observing from Madras (today known as Chennai) fared no better.

Still, the problems continued as Le Gentil made his way back to France. On the way back, he had to face shipwrecking twice. When he landed in France, he discovered that since he had been out of touch with his friends and relations, he had been declared 'legally dead' and all his property given to legal heirs.

The story ends with a happy event, however. Le Gentil proved himself alive and got his property back, followed by a happy marriage.

In this story, we get some idea of problems faced by European astronomers visiting India for observations. The trials and tribulations that Le Gentil had to face when he undertook the visit to India for observing the transit of Venus were indeed exceptional but not altogether uncommon. The British also had deputed an observer for the same purpose, but there is no record of what calamities he had to face. But we may very well raise the question: "Was any native Indian observing that event?" The answer is: "Most probably there was none." Certainly, none from the leading Indian astronomers at that time. This raises an additional issue: "Did no one in India feel the urge to observe?" Indeed, if that were the situation, it reflects on the lukewarm interest in science (which includes astronomy!) in the country at the time. And how can science flourish if there is a basic lack of interest in science?

Thus, we have tried to answer Salam's question and hope we can learn from the answers!

Needham's question

Like Salam's question about the Indian subcontinent, there was a similar question about China raised by another scientist in the twentieth century.

Joseph Needham (1900–1995) from Cambridge University, a scientist by origin, got interested in China as a country and while commenting on various cultural aspects of the country, expressed surprise at the lack of growth of science in that country. While raising the question, Needham tried to provide the answer somewhat along the line we discussed for India.

One reason that arose was that the Chinese kings were involved in wars. Every new emperor felt obliged to engage in warfare, believing that it was not only obligatory but also desirable. This belief had no place either for the generation of new ideas or for a proper allocation of money for research.

Another reason was of a social kind. The most respected profession was government service. So, when a bright young man, after doing well in basic education, looked for a suitable respected profession for higher training, science as an option was not available to him. Thus there was no way innovation could flourish. Indeed, such youths were positively discouraged from trying out something new. One can contrast the European and Chinese situations for a reason.

We conclude this discussion by noting that a considerable controversy exists and Needham himself had to defend the answers to his question.

5

GEOMETRY AND ALGEBRA COMBINED

The birth of new ideas

The end of the sixteenth century heralded many changes in Europe. A list of some famous persons in different fields indicates that creativity was exploding in many different forms. To name a few, in mathematics and astronomy, we have Copernicus and Galileo. In the literary field, there was the dramatist William Shakespeare and the poet Milton. In the medical field, one could mention Harvey, the discoverer of blood circulation. Amongst scientists, early electromagnetic experimentalist Gilbert and amongst mathematicians Pascal, Fermat and Descartes could be mentioned.

In those early days of mathematics, the beginning was made with arithmetic – telling man how numbers are necessary and useful in dealing with many day-to-day matters. Euclid added the new branch of mathematics illustrating the use of figures. In due course, the use of algebra showed how to enlarge the scope of arithmetic.

Rene Descartes

A similar union of algebra with geometry was expected but needed a perceptive mathematician to make a beginning. Rene Descartes was the person to do so. The so-called coordinate geometry was the result of the union. The Figure 5.1 demonstrates the use of this new branch of mathematics.

The originator of this new branch of mathematics was born on 31 March 1596, in the French town of La Haye near the bigger city of tours. The period of Descartes's life was at a time when there were major social changes in the offing. There were new ideas, new social habits, new religious perceptions trying to replace the earlier view. Naturally, there were conflicts, illnesses arising from imperfect social hygiene and public controversies about social, religious and political issues. Descartes was

DOI: 10.4324/9781003203100-6

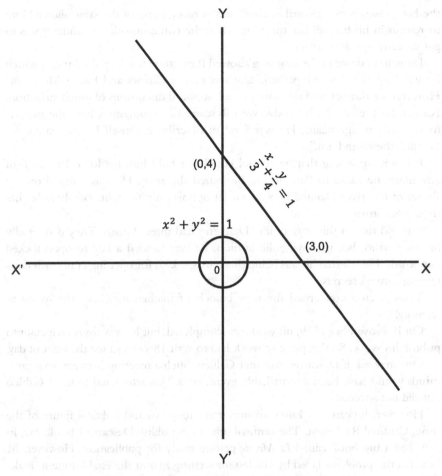

$(0,4)$

$$\frac{x}{3} + \frac{y}{4} = 1$$

$$x^2 + y^2 = 1$$

$(3,0)$

X'

X

Y'

0

Cartesian/Coordinate Geometry

FIGURE 5.1 Cartesian geometry. This figure demonstrates the basic rules of Cartesian geometry. The plane is marked by two (X and Y) axes and geometrical figures are represented by algebraic equations. Two typical examples (straight line and circle) are shown.

born in a wealthy upper-class family, but he was unfortunate to lose his mother (Jeanne Brochard) shortly after his birth.

His father, however, made sure that Rene and his two elder brothers received a good upbringing. Rene himself had delicate health and learnt to avoid 'physical interaction.' He was always concerned with matters related to the head and brain. Indeed at the age of 8, the boy used to raise questions and argue with his caretaker. However, things improved when Rene was admitted to the residential school La Fleche. The rector of the school, Father Charlet, soon discovered that the boy possesses exceptional intellect and in recognition of the same allowed him to remain in his bed till late morning when the common rule for students was to get up early in the morning.

Lying in bed late in the morning allowed Rene time for deep thinking in which he was helped by his sympathetic teachers Father Charlet and Father Mersenne. However, he did not find religious or philosophical discussions of much attraction. Rather, he preferred what today we call scientific arguments where the experimental role is significant. He was fond of describing himself by the statement, "I think therefore I am."

It is not surprising that his school could not hold him for long. In search of adventure, he came to Paris where he joined the army. He was camped on the shores of the river Danube hoping for an opportunity to fight. But here his life took a new turn.

It is said that in this state of life, Descartes had three dreams. They had rapidly blowing wind, but in a symbolic fashion, he was handed a key to open locked cupboards. He felt that he was being given magic keys for opening cupboards containing nature's secrets.

Those secrets concerned the new branch of mathematics, viz. the *coordinate geometry*!

On 10 November 1619, his work was completed, but he was always reluctant to publish his work. So this piece of work had to wait 18 years to see the light of day.

One may ask if Descartes ever met Galileo. Such a meeting between two great minds would have been a remarkable event. Alas, Descartes tried to meet Galileo but did not succeed.

However, he came to know an important religious and political figure of the time: Cardinal Richelieu. The cardinal offered to publish Descartes's book. He, in fact, had a big book called *Le Monde* getting ready for publication. However, he heard of the problems faced by Galileo for writing against the establishment, making fun at its expense. As he himself had ridiculed the establishment, he was too scared to publish. He left instructions to publish his work posthumously, except for one book called *The Method* which was published in 1637. With the cardinal himself backing him, he had nothing to fear in France.

Coordinate geometry

The method employed by Rene Descartes was to express a geometrical statement in terms of equations and variables x and y defined as follows.

First, we have two straight lines, called the *axes*, in perpendicular directions and intersecting in a point we denote by O. (Refer to the Figure 5.1 drawn for convenience.) Denote an arbitrary point P by the two 'coordinates' $[x, y]$ as follows. Draw perpendiculars PX and PY from P on the two lines. Denote the distances PX and PY by the variables x and y. In case the point P lies below the relevant line, we attach a negative value to x. Likewise, we attach a negative value to the variable y if the point P lies to the left of the relevant line. The basic lines are called x and y axes.

The numbers called 'coordinates' are identified in this way to define a typical point on the plane. To make a transformation to geometry, we proceed as follows. Suppose we consider all points that lie in this plane such that their coordinates satisfy the relation

$$x/a + y/b = 1,$$

where a and b are given numbers. What does this condition tell us? It is easy to verify that the description is of a straight line which has an intercept a on the x axis and b on the y axis.

Likewise the relation

$$X^2 + y^2 = R^2$$

denotes a circle of radius R having its centre at the 'origin' – that is, the point with $x = 0$ $y = 0$.

Thus, we can use the previous definitions to solve geometrical problems. Clearly, any geometrical statement can be translated into algebraic equations.

As against the Catholic support, when Descartes went to settle down in Holland, he was violently opposed by the Protestant intelligentsia. However, since he had a friend and supporter in the ruler, the Prince of Orange, he had nothing to fear.

Pierre de Fermat

A contemporary of Descartes was Pierre de Fermat, a brilliant mathematician who looked upon mathematics not as a means of solving daily problems but as a source of fun. He had a relatively steady life in a government job with no great controversies to face. He was 'commissioner of requests' in the town of Toulouse from the age of 30, and till his death at the age of 65, he continued in government service.

Although Fermat was more interested in aesthetic aspects of mathematics, some of his work was of practical use also. For example, Fermat found the rule of constructing the tangent to any given curve, using the same prescription as found by Descartes. Also, he found a general rule for transmission of light between two fixed points with reflections and refractions en route (see Figure 5.1).

Apart from this, Fermat is best known for his work on number theory. For example, his work on prime numbers is well-known and understood. A number is prime if it is not divisible by any number other than itself and by 1. Thus the sequence 2, 3, 5, 7, 11, 13, . . . is a sequence of primes. Now consider the following numbers:

$$2 + 1, \ 2^2 + 1, \ 2^4 + 1, \ 2^8 + 1, \ . . . \ 2^{16} + 1. \ . . .$$

If we work these out, we get, respectively, 3, 5, 17, 257, 65537. . . and we can further verify that they are all prime. Based on this, Fermat had a conjecture that all numbers expressible as

$$2^{2 \times 2 \times 2 \times 2 \times n \text{ times}} + 1$$

are prime. But he was wrong! For $n = 5$, we get 4294967297, which is not a prime, being divisible by 641. But on primes, Fermat did find a good theorem called Fermat's little theorem, which is as follows. If n is a whole number and p is a prime then $n^p - n$ is divisible by p.

For example, for $n = 2$, $p = 7$, we have $n^b - n = 126$, which is divisible by 7.

Fermat had a reputation for making exact statements. If he said that he had proved a theorem he would be believed by his colleagues since he would be able to give the proof. This credibility led to a long-standing problem as follows.

We know from Pythagoras's theorem of geometry that the square of the hypotenuse of a right-angle triangle equals the sum of squares of the two sides. Thus for example, the three sides can be of lengths 3, 4, 5. In general, we can find any number of these 'Pythagorean' triplets of whole numbers. In fact, by writing $a = m^2 - n^2$, $b = 2mn$ and $c = m^2 + n^2$, where m and n are whole numbers, we can ensure that the Pythagorean relation is satisfied with

$$a^2 + b^2 = c^2.$$

With this background, Fermat raised the question, Is it possible to find triplets of whole numbers satisfying the relation

$$a^3 + b^3 = c^3$$

or a triplet in which the power is not 3 but any higher index higher than 2? Fermat was reading the book by Bachet on Diophentus and had scribbled in the margin that he had an elegant proof of a theorem on this issue but could not give the proof of it in the margin of the book he was reading. The theorem itself can be stated by saying that such triplets of whole numbers cannot exist for any index $n > 2$.

Fermat's note sent a lot of mathematicians in search of the proof of *Fermat's last theorem*. We will return to this issue in Epilogue.

Blaise Pascal: a new kind of geometry?

Fermat and Descartes had another distinguished contemporary: Blaise Pascal. Pascal was also French, born in Clairemont on 19 June 1623. His father, Etienne Pascal, was in government service and well-known for his knowledge and wisdom. He was unfortunate to lose his mother when he was only 4 years old. But his father took care that the child did not suffer in any way. Besides, he had close interactions with his sisters Gilberte (Mme Perier after her marriage) and Jacqueline who later took holy orders to become a nun.

Actually, Pascal came to be recognized for his writings not in mathematics but for their literary quality. Known as "Thoughts" and "Provincial Letters," these writings were well recognized in the French literary sphere, but Pascal's real favourite was mathematics. Indeed he enjoyed Euclid's book, which his father had gotten for him. In fact, his father was very pleased when the boy Pascal independently proved the theorem that the three angles of any triangle add up to two right angles.

Pascal continued to improve in his mathematics, and because of his ability in mathematics, he was allowed to attend Father Mersenne's (the same one who had taught Descartes) discussion group meetings. However, because of his forthrightness in expressing his opinion, he was not popular and had to suffer for getting into the bad graces of Cardinal Richlieu. Indeed, to escape punishment, which the politically powerful cardinal was capable of inflicting, Pascal had to live incognito with his extended family. However, the cardinal happened to watch a play in which Pascal's sister (Jacqueline) had acted superbly. The cardinal was greatly impressed. Learning that she belonged to the Pascal family, her acting performance made the cardinal change his mind, and thereafter he was no longer hostile.

So at last the Pascal family returned to Paris and could live there peacefully. Pascal's real interest was in mathematics to which he could devote good time. Now that he was in Paris where his contemporary Descartes also lived, the question arises as to whether they met and discussed the subject. They did meet and discuss but not without fighting about mathematical issues. In particular, Descartes could not believe that the famous book *Essai pour les conics* by Pascal was written by him at the age of 16. Likewise, Descartes believed that some of his experiments on air pressure had been plagiarized by Pascal. Since both had attended Father Mersenne's seminars, Descartes had suspected that in one of the sessions, Pascal had attended his presentation and so had copied his experiments there. Thus, there were acrimonious meetings between the two geniuses!

However, there was one positive suggestion that Descartes made to Pascal: "Spend a lot of time in bed after you wake up in the morning."

Unfortunately, Pascal did not enjoy good health and a long life. He suffered from dyspepsia and insomnia for 16 years. There were several new discoveries he would have contributed to if he'd had freedom from the health problems. But that was not to be; he died at the young age of 31. He is also credited with the discovery of the calculating machine.

He found a new way of describing geometry, called *projective geometry*. Many of the results in Euclid's geometry could be augmented by the clever technique Descartes used. See the adjoining that gives glimpses into his work.

Projective geometry

This new type of geometry was used by Blaise Pascal to arrive at new geometrical results by a clever transformation. Known as projective geometry, this approach provides the development of geometry through projection. The shadow on the ground or cine projection on the wall provides examples of how shapes change under projection.

Take the example of parallel lines extending all the way up to infinity. However, if we 'look' at them from one end (see Figure 5.2), we see them apparently converging so that we can imagine them meeting at a 'point at infinity.' Thus mathematically, we may imagine a point at infinity and denote it by Ω. It turns out that several theorems in geometry not having to do with absolute lengths or angles can be taken over for systems having such points at infinity.

FIGURE 5.2 Projective geometry explained. It deals with projections at large distances (ideally 'infinity'). The railway track is made of parallel lines but appears to come out of a point at infinity.

6

THE SUN RISES

Introduction

The arrival on the scene of Isaac Newton spelt a fundamental change in the status of mathematics and science. So far as science was concerned, the notion of giving a quantitative measurement when describing a scientific event really caught on after Newton had demonstrated specific examples. Thus, the event of an apple falling on Newton's head (an event which may or may not have happened) may have glamorized it but in actuality, Newton went by Kepler's interpretation of Tycho Brahe's data.

Mathematics also was feeling the need to expand well beyond the classical trio of arithmetic, algebra and geometry. Indeed, during the late seventeenth century, the discovery of calculus by Newton and Leibnitz opened new passageways for new branches of mathematics.

For this reason, both in mathematics and science, we could describe the development as *sunrise*.

Isaac Newton: the early days

Isaac Newton was born in the Woolsthorpe Manor House in Lincolnshire on Christmas Day, 1642 by the old (Julian) calendar, which would correspond to 4 January 1643 by the reformed Gregorian calendar. It is pointed out in several places that Newton was born in the year Galileo died. Nothing should be read into this match of dates! Unless you are a believer in the Hindu concept of rebirth! Newton's father died three months before Newton was born. His mother remarried. But the child Isaac (as Newton was named) did not get along with the stepfather. As a result, he was cared for by his grandmother (mother's mother) and so was closer to his grandmother and maternal uncle than to his mother. It was his maternal uncle

DOI: 10.4324/9781003203100-7

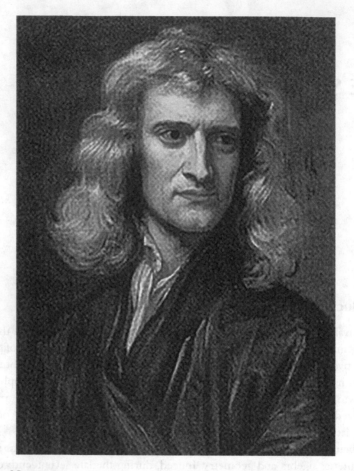

PHOTO: Newton

William Ayscough who spotted sparks of intelligence in the growing youth and took the important step of sending him to Cambridge for higher studies.

In those days Cambridge had three types of students. There were the scholars who managed on their scholarships; then there were the well-off students who managed on their personal family income, and the last category (to which Newton belonged), the *subsizars* were students who earned their keep through menial work. Newton had registered for a degree in law but later shifted to science and mathematics as his interest was captured by those subjects and their teachers. And although he took the B.A. degree in these subjects, his brain was more stimulated by the problems of research. In 1661, he had earned admission in the lowest category mentioned previously, but in 1664, he had progressed to the scholar category.

A fortuitous circumstance then really transformed him into a brilliant researcher. Because of a nationwide plague epidemic, the university was closed, and all students were asked to go home. Thus Newton found himself in Woolsthorpe, all isolated with a lot of time to think. The two years that he had in this capacity led him to brilliant creativity.

Limitations of space and time do not allow us to go into detail about Newton's achievements during the years 1665–1666. We will do a superficial job here: our main aim being to show the way modern science owes its origin to Newton.

The laws of motion

The foundations of dynamics lie in Newton's three laws of motion of which the first law was earlier arrived at by Galileo. Known sometimes as the law of inertia, it maintains that for anybody the state (of rest or uniform motion) does not change unless some external force acts on it.

Till this law became understood and established, the dictum of Aristotle held sway, which argued that motion requires force. In his book, Galileo had given the counter-example of an arrow. Why does it move in the air if there is manifestly no force acting on it? To this, the Aristotelian would reply that the arrow moves forward because air pushes it. In that case, Galileo asked why the arrow does not go in the lateral direction since that way the air has more area of the arrow to push against.

The effect of force as the agency changing the state of motion follows in the second law of motion. And finally comes the third law equating action with reaction. These are perhaps the first set of laws of science but their universality is remarkable.

What makes an arrow, shot from a bow, travel?

Those brought up on the ideas of Aristotle firmly believed that force caused by an agency is required to keep a body moving. To support the reasoning, they gave the example of a pushcart which keeps moving so long as there is someone pushing. The moment the push is stopped, the cart stops moving.

Galileo countered this reasoning, first by giving the example of an arrow shot from a bow (Figure 6.1). As the arrow keeps moving in the air, where is the force acting on it, he asked. The Aristotelians replied that the air in which the arrow moves pushes the arrow forward. To this, Galileo replied by saying that assuming such a push exists, it should be more on the arrow if it is shot laterally as shown in the figure. Because an arrow shot lengthwise has a smaller pushing area than an arrow shot laterally, we expect the laterally shot arrow will go much farther than an arrow shot lengthwise. This does not happen. The arrow shot laterally does not get very far as seen in the lower figure.

The pushcart is acted on by friction which causes it to stop when the pushing force is withdrawn. But *it does not stop instantly*. It slows down and then comes to a halt. This shows that friction causes the cart to decelerate until it moves with zero speed; that is, it comes to a halt. If Aristotle were right, the cart should have stopped instantly when pushing was halted.

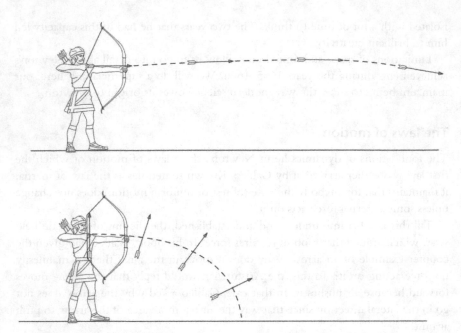

FIGURE 6.1 Galileo used the example of shooting an arrow to debunk the Aristotelian concept of motion.

In Newton's second law of motion, the key word is acceleration and not velocity. Thus in the case of the pushcart, it is to be noted that if the pushing is stopped, the cart does not come to a halt instantly. But it decelerates and eventually comes to a halt.

The law of gravitation

This is the first law of basic physics describing interaction between material objects. Newton has not commented anywhere on the way he came to think of it. Indeed he used his own discovery of calculus to solve the basic question, What must be the force of gravitational attraction of the Sun on a planet so that the latter moves as per Kepler's laws? The answer that the force must be "inverse square" type is a fitting demonstration of how the laws of motion operate.

There are in literature claims ascribing the discovery of the law of gravitation to others. For example, Bhaskaracharya is stated to have argued that things fall on the earth because of its force of attraction. While such a claim might be considered to be on the right lines, it is at best qualitative and falls well short of any quantitative prediction. In Newton's own times, there was the claim by Robert Hooke who also attacked Newton's work on optics. Whatever the basis of others' claims, the claim that the demonstration by Newton of how Kepler's laws were explained by

PHOTO: Hooke

the laws of motion together with the inverse square law does not seem to have been provided by any other claimants.

Invention of calculus

In order to demonstrate many results in science like Kepler's laws, as well as in basic mathematics, Newton began to use another discovery of his – namely, the branch of mathematics called calculus. There were many results that looked cumbersome to derive but became greatly simplified by using this new technique. Even so, feeling that his contemporaries would not understand (and therefore mistrust) the calculus-based proofs, he re-derived them using the then conventional techniques based on algebra and geometry.

Experiments in optics

Newton showed how the sunlight passing through a prism splits into seven colours: violet, indigo, blue, green, yellow, orange and red (the colours are often remembered through the acronym VIBGYOR). This was the beginning of the science of light which was to take more than two centuries for a full understanding. But in the early stage, Newton saw that refraction of light by passing from one medium to another breaks up into these colours. This phenomenon makes an image by refraction blurred. This defect is called *chromatic aberration* (Figure 6.2).

Chromatic aberration

Newton knew from his various experiments in optics that a light ray changes its direction when it crosses over from one medium to another. He had also found that this property of *refraction* is different for light of different colours. As shown in Figure 6.2, light passing through a prism is bent more for violet colour and is the least for red colour.

This property of light causes a practical difficulty to the astronomer while building a telescope. Imagine that the telescope is looking at a star image. The light from the star will be split in seven colours and each colour will form an image separately. Thus instead of a clear distinct image with original starlight, the observer gets a

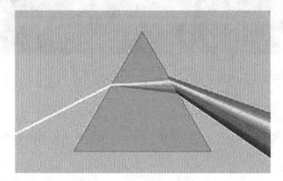

FIGURE 6.2   Prism experiment. Newton used a prism to show that normal sunlight is made of seven colours. While passing through the prism, the sunlight gets refracted thus changing its direction. This bending is more for violet and less for red. This causes the seven colours to separate. The telescope using refracting lens, therefore, causes seven coloured images to form with slight displacement as shown in the figure. This phenomenon is called chromatic aberration, and it requires a 'correcting lens.'

fuzzy image formed by the superposition of the seven coloured images. This defect is known as *chromatic aberration*. Although the defect can be 'corrected' by inserting more lenses in the telescope, the correction is not perfect.

Realizing this defect, Newton decided upon a different type of telescope in which the bending of light is carried out by a concave mirror. When light is bent through reflection, its path is the same regardless of colour. So the Newtonian model of the telescope is relatively easier to make.

Newton is credited with a new model of telescope, one which in a modified form turned out to be superior to the then commonly used Galilean version. The main difference was that instead of a principal lens, the Newtonian telescope had a principal *concave mirror*. As seen in Figure 6.2, this made the telescope more compact. Also, the fault of chromatic aberration is absent in this model.

These are highlights of Newton's creativity in the two productive years 1665–1666. When he returned to Cambridge, he was elected fellow of his college, Trinity, in 1667, and two years later, on the recommendation of his former teacher of mathematics, Isaac Barrow, he was given the Lucasian Professorship, which Barrow himself had held. We will not go into detail about Newton at Trinity but make a few topical remarks.

Controversies in Newton's life

The name of Isaac Newton is associated with so many important contributions that it is not surprising that he would be involved in various controversies. Here are some which generated considerable heat.

It is generally assumed that Newton had discovered the inverse square law of gravitation back in 1666 when he was away from Cambridge and working at his home. He had shown that with such a law of attraction, the planets would move in elliptical orbits as empirically shown by Kepler. This was a major achievement, but Newton did not publish this finding at that time. Subsequently, in 1684, Newton was asked by his friend Edmond Halley as to what law of force between the Sun and the planet would give rise to an elliptical orbit for the planet. To this Newton replied, "It has to be the inverse square law." On Halley asking on what basis Newton had so claimed, the latter replied that he had worked it out.

Now, Halley had probed Newton on this point because distinguished persons like Sir Christopher Wren had been anxious to know the real situation in view of a controversy started by Robert Hooke. Hooke claimed to have the authorship of the law of gravitation and in that sense, there was an accusation of plagiarism against Newton.

So when Halley asked the aforementioned question, in order to substantiate his claim, Newton looked for his 1666 work but could not find it. So he started to derive the above result. But somewhere he made a mistake. However, he soon found it and could confirm the result. But why did he wait 18 years before publishing it? The reason given is as follows.

Assuming that the inverse square law applies to point masses, a mass M will produce a force GM/R^2 at a distance R from it. Real masses are not points but extended objects. So how does a spherical mass of a given radius attract? This was a non-trivial question, one for which there was no answer available. It is said that Newton waited till he could find the answer, which required the application of integral calculus. This caused the delay in his getting the result. Finally, when he could show that a spherical object attracts as if all its mass is located at the centre, he was confident of his theory. It is also argued that lunar tables showing the Moon's motion were under preparation and took a longer time, and Newton wanted to check his theory against those data. This is why when the controversy erupted because of his describing the problem of planetary motion in the first two parts of his *magnum opus* the *Principia*, he wrote to Halley expressing his disgust with the controversy: "*The third (part of his book) I now design to suppress. Philosophy is such an impertinently litigious lady that a man had as good be engaged in law suits to have to do with her.*"

How do spheres attract?

Newton was a perfectionist so far as his expectations from the laws were concerned. When he wrote the 'inverse square law,' the masses it talked about were point-shaped. However, when he started applying the law, he was in trouble. If the Sun and Earth are supposed to attract each other, do they follow the rule that the spherical shapes make no difference to the answer? To show this to his satisfaction, Newton wanted to prove it mathematically. This took him some time as he used his newly found calculus for the proof.

It is said that Newton waited till he sorted this point before publishing his inverse square law. The delay in publication, however, led to misunderstanding and more controversies about the authorship of the law of gravitation.

Whatever the case, it is quite clear that the mathematical derivation of planetary motion presented a problem which only Newton could solve given his mathematical ability. Still, the fact that his authorship of the above law was questioned led him to a decision to not publish his further work. Fortunately for posterity, Halley persuaded Newton to lift his self-imposed ban on further publication.

While this controversy between Hooke and Newton was confined to England, his controversy with the German scientist Leibnitz spread all over Europe. The issue this time was: who invented the mathematical branch of calculus? The European view was that the credit went to Leibnitz, whereas the English belief gave authorship credit to Newton. Each side supported its champion while also claiming that the other champion had used the results without giving due credit to the opponent. Going beyond mathematical issues, Johann Bernoulli wrote an open letter in which he made several accusations against Newton. This letter got wide publicity and brought the interaction between England and Europe under considerable strain, although Bernoulli later denied that he had written such a letter. Newton himself reacted by saying, "*I never worked for international fame but I have always been conscious about keeping my probity intact. The letter writer has sought to deprive me of it and taken on the role of a judge. It is my practice not to get into arguments to defend my views.*"

Not to be outdone, in England, a commission was appointed by the Royal Society to investigate the truth of the matter. This committee heard only Newton's side and unilaterally decided that Newton was the true originator of calculus and that Leibnitz had copied him. A further ridiculous aspect of this farce was that the committee's report was written by Newton himself!

Wave or particles?

The basic understanding of light was still far from perfect in Newton's time. Two alternatives were available. Either the light was a collection of a group of particles, or it had a wavelike structure. Newton supported a mechanistic point of view and so strongly believed in the former theory popularly known as the *corpuscular theory*. Such was the force of his personality that the physicists in Britain could not go far to look at the alternative explanation.

The alternative of *wave theory* flourished on the continent, however. Hyghens from Holland was the main scientist to work on the wave theory of light. Phenomena like refraction, interference and polarization could be nicely explained by wave theory but had difficulty with the corpuscular theory.

However, as we shall see later, the quantum theory gave a comprehensive explanation of particle and wave theory of light.

Newton's work on optics – that is, on light rays – while very perceptive, also led him to another controversy. Is a light ray made of a large number of particles, thus giving the theory of light the name 'corpuscular' or is it made of waves? Newton, who had supported the former theory, had done interesting experimental work on light. Thus he was able to show that sunlight is made of the superposition of seven colours (violet, indigo, blue, green, yellow, orange and red – thus making the acronym VIBGYOR). He had studied reflection and refraction of light and used his experimental skill to design a reflecting telescope which in due course replaced the refracting telescope used by Galileo.

But experimentalists in other countries like Huygens in Holland had been converging on the latter of the two theories and thereby ran into powerful opposition. For Newton's views were hard to override and one could say that because of Newton's opposition, progress in the understanding of light was delayed. Although he stuck to the corpuscular theory, Newton was aware of certain experiments that were difficult to understand based on his hypothesis and which could be explained by the wave theory. But he was publicly against it. Although later developments of quantum theory brought back the particle nature by introducing the notion of the 'photon,' the basic concept was different from Newton's, as we shall see later.

Newton's query

When Isaac Newton formulated his laws of motion and gravitation, he was faced with one fundamental question: is light affected by gravity? Recall that when Newton was formulating the law of gravitation, he called it the law of *universal* gravitation. That meant that any two masses in the universe should attract each other. So does the universality cover light also?

In practical terms, this meant the following. If a beam of light is passing by a lump of matter, will the mass of the lump attract the light beam and bend it as shown in Figure 6.3? If the light is attracted by gravity, the answer to this question is 'yes'; otherwise, it is 'no.' What was Newton's answer to this question?

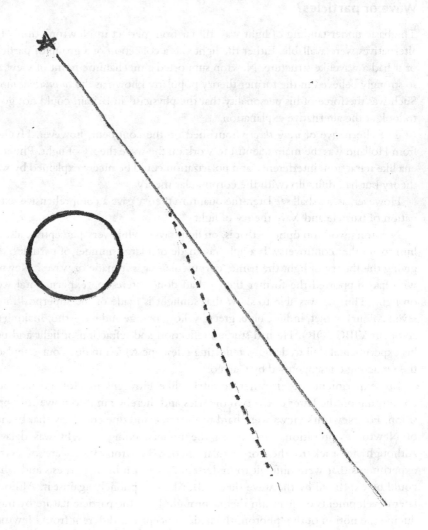

FIGURE 6.3 Bending of light. A ray of light passing close to a massive object. Newton's query was whether it bends or goes straight.

Newton, however, took a strong line on the general issue of scientific fact. So far as any scientific statement was concerned, he refused to admit it as factual, unless it was backed by facts such as an experimental verification or an observed fact. Once when he was asked his opinion on a certain basic issue of science, he declined to offer it saying, "*Non fingo hypothesis*," meaning, I do not frame ideas that are pure guesswork. Indeed, so far as light was concerned, his book entitled *Opticks* contained such guesswork ideas in the form of *queries*. The very first query concerns the bending of light by gravity:

> *Do not Bodies act upon Light at a distance, and by their action bend its Rays; and is not this action [caeteris paribus] strongest at the least distance?*

Bending of light by Newtonian gravity

Although Newton would not be drawn into a discussion on whether his inverse square law did predict a gravitational effect on light rays, later physicists tried to work out the effect if present. The first to do a complete job was Johann Georg von Soldner in 1804. Before coming to calculations, we need to make certain new assumptions to supplement the standard assumptions made by Newton.

Thus, we assume that a light ray has photon-like particles, and they travel with speeds comparable to c, the 'speed of light in vacuum.' Also, the photons otherwise are very small particles subject to Newton's laws of motion and gravitation.

The light-like particles going near a massive object M are attracted by the gravity of the mass. Since the motion of the photon is very large, its trajectory is hyperbolic. Two asymptotes of the hyperbola are then computed using the trajectory noted earlier. The angle between the asymptotes is very small, but it is the effect we are looking for.

As became well-known, the effect predicted by this hybrid Newtonian calculation is half of that Albert Einstein calculated for his general theory of relativity. The Newtonian prediction for bending produced by a spherical mass M having radius R in a ray grazing its surface is $2GM/c^2R$. The relativistic result is twice this value as we shall see later.

7

SCIENCE AFTER NEWTON

As we saw in the case of mathematics, Newton's work in science also inspired the progress of various other studies. We will now briefly review the important developments inspired by Newton's scientific work. But it is better first to eliminate from it all his pseudo-scientific work. Following the superstitious beliefs in those times, many practitioners of science indulged in subjects like alchemy and astrology.

From alchemy to chemistry

The science of chemistry was preceded by the pseudo-science of alchemy. It was popular on the European subcontinent. Its popularity was based on various unsubstantiated claims, such as providing a process for converting mercury to gold and having a magic wand in the form of the philosopher's stone. But realizing that such claims are all hoaxes, the government in Britain had placed a ban on alchemy.

Thus, practitioners of alchemy, if they wished to conduct experiments, were forced to do it on the sly. Even a great scientist like Newton had a secret lab where he carried out his 'illegal' studies of alchemy. But of course, nothing came out of that effort. Likewise, there is some talk that Newton also believed in astrology but there appears to be no evidence of it.

The beginning of the science of chemistry can be traced to the earlier interest in alchemy. In France, the scientist Lavoisier introduced the concept of experiments based on scientific procedures with due records kept of different trials. This was the beginning of the science of chemistry. Measuring chemical reactions, Lavoisier could show that the total mass of the participating components of a chemical reaction equals the total mass of the products coming out of it. To show this, he made a balance of high accuracy.

Unfortunately, the time was of great upheaval in France after the revolution, and several innocent persons were sentenced to death by the guillotine, and Lavoisier

DOI: 10.4324/9781003203100-8

was one of them. But the momentum generated by his work was sufficient to lead chemistry on to further progress. We will encounter some landmarks in due course.

Inputs to astronomy

Side by side with terrestrial science, Newton had contributed to astronomical science as well. Here too can be seen his all-around ability – theoretical as well as experimental.

The deduction of the inverse square law of gravitation from Kepler's laws was a stroke of genius. It demonstrated how a scientific law can account for dynamical behaviour on a cosmic scale. Not only that but also in 1683, Newton's friend Edmund Halley made the remarkable deduction that a series of cometary visits almost every 76 years indicated that it could be one comet periodically visiting every 76 years. Halley had based his conjecture on the assumption that the Sun attracted the comet in much the same way that it attracted planets. To put his deduction to the test, Halley made the prediction that the same comet would be seen in the year 1757–1758. This prediction was duly borne out, although Halley himself was not alive then.

Later, as happened in the case of planet Uranus in the 1840s, the apparently irregular motion of that planet was caused by the perturbation from an as yet undiscovered planet. This deduction and its verification led to the finding of the planet now known as *Neptune*.

The discovery of Neptune

It is said that a scientific theory is never proved but at some stage, new tests might disprove it. That means that when a theory makes a prediction, future tests may either confirm it or disprove it. Thus at no stage can one be sure that the theory is correct since a new test may come along to disprove it. The history of the law of gravitation bears this out.

Newton's law of gravitation successfully explained the motions of planets and satellites in the solar system. It successfully predicted the arrival day of Halley's Comet. Yet, there came a day when another test threatened its validity. The newly discovered planet Uranus did not appear to move as per Newtonian law. Careful observations of its motion confirmed this suspicion.

Did the observations show a genuine discrepancy implying an error in Newton's laws of motion and gravitation?

Two young astronomers, John Couch Adams in Cambridge and Urbaine J. J. Leverrier in Paris independently came to the conclusion that there was no problem with the laws. What was missing was a hidden planet whose gravitational field was interfering with the motion of Uranus. Both astronomers arrived at the same conclusion about the location and motion of the new planet and called upon the senior astronomers in their countries to look for it and check their predictions.

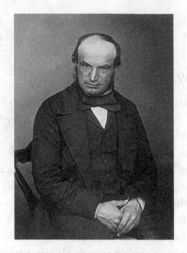

PHOTOS: Adams, Leverrier

Much drama has been written on this episode. The bottom line is that senior astronomers in both England and France did not take the prediction seriously, and, finally, the German astronomer Galle, a young man in the Berlin Observatory who happened to be holding the fort while his boss was away celebrating his birthday, took the prediction seriously and found the new planet as predicted. Thus was added an extra planet to the solar system.

PHOTO: Galle

However, that was not the end of the story, as we shall find out later!

Leaving aside these examples of how theoretical attempts to apply Newtonian ideas led to further successes, we now highlight the observational fallouts of Newton's ideas. We begin with the Newtonian model of the astronomical telescope: for Newton's most important practical input was in the making of astronomical telescopes. The model propagated by Galileo had several shortcomings which the Newtonian version sought to remove. We describe a few points to illustrate the situation. For this, refer to the adjoining figure of both Galilean and Newtonian models (Figure 7.1).

1   We note that the Galilean model had a main lens for collecting and focussing the rays coming from the object, which is a distant source of light. The Newtonian model has a concave mirror doing the same job. However, as we start building a large telescope, we discover that making a concave mirror is technically easier than making a lens.

2   A lens has the object on one side and the viewer on the opposite side with the result that the telescope has to be very long. The Newtonian model has both object and the viewer on the same side so that the model is compact and easy to install or carry.

FIGURE 7.1   Telescopes of Galileo and Newton. Newton's model is compact and does not suffer from chromatic aberration.

3   The lens is made of some transparent material and will cause partial absorption of light going through. The reflecting model does not have this defect.
4   The refracted light can show different wavelengths bending differently thus making the images less sharp. We referred to this 'chromatic aberration' in the last chapter. In the reflecting (Newtonian) model, this defect is not present.

It is not surprising that today's big telescopes use reflection and not refraction as the means of collecting light from an astronomical source.

There is a saying in India that other trees do not grow under the shade of the banyan 1tree. A banyan tree is vast in its spread and any smaller tree underneath it does not get sufficient sunlight to grow fully.

During Newton's lifetime, there was progress in science and mathematics but rather differently in Britain and in Europe. For example, the Newton-Leibnitz controversy led to the development of pure mathematics in Europe whereas The use of calculus and other branches of mathematics in solving problems of applied mathematics caught on more in Britain.

In some respects, Newtonian prejudices positively prevented progress in certain branches of science. For example, the wave theory of light was rejected by Newton in favour of his own corpuscular theory. It took several years after Newton for progress to resume. On the other hand, we have cited the example of the Bernoullis who were responsible for several practical developments. At the same time, there were continental mathematicians like Euler, Lagrange, Fourier, etc., who led to the growth of several new branches of mathematics. We will briefly describe them here, leaving the growth of the applied aspects to the next chapter.

The Bernoullis

Of the earlier Bernoullis brothers, Jacob and Johann were from Basle. Ignoring their father's wish that they continue in the family business of spices and failing that have a career in the church, they opted for mathematical careers. Both became famous, although there was rivalry between the two. In particular, Johann felt jealous of Jacob for his position as Professor at Basle University. This led often to exchanges of bad words or not giving proper credit when due. This lasted till Jacob died and Johann was promoted to his post at the university. The family tree of Bernoullis starting with Nicolaus is shown here.

Mathematicians Bernoulli family tree

Nicolaus (senior) (1623–1708)
Jacob I Nicolaus I Johann I
1654–1705 1662–1716 1667–1748
Nicolaus II Nicolaus III Daniel Johann II
1687–1759 1695–1726 1700–1782 1710–1790
Johann III Jacob II
1746–1807 1759–1769

One of Johann's students who later became famous was Euler, whose work we will describe shortly. But Daniel, his son, also became famous for his work. Johann then became jealous of his son and resorted to unethical practices, like plagiarizing Daniel's work by announcing it as his own, writing a book on Daniel's work but under his own authorship, etc. Later, Johann himself was at the receiving end when his student L'Hopital wrote under his own name a book based on Johann's work. The well-known L'Hopital's Rule was really the finding of Johann. The rule shows a simple way of performing limiting operations.

It will take us too far into technical details to describe the works of all the Bernoullis. At the end is described the challenge problem posed by Johann Bernoulli, which was solved by Newton. The following anecdote will indicate how well-known they had become in Europe because of their work.

Daniel Bernoulli had become famous at a young age. Once he was travelling on a long-distance coach with a fellow traveller. Daniel introduced himself quite simply: "I am Daniel Bernoulli." His fellow traveller could not believe that a man so young could have achieved so much and thought that he was exaggerating.

To show his disbelief, he replied, "And I am Isaac Newton." This was the best compliment for Daniel as he himself felt.

The challenge problem posed by Johann Bernoulli

The growth and progress of mathematics are owed in no small measure to the occasional difficult problems posed by working mathematicians to their colleagues. Fermat's last theorem described on page 40 was one such problem. Of course, it took an exceptionally long time to solve, but many problems get solved in a shorter period.

Posing a 'challenge problem' to the contemporary European colleagues, Johann Bernoulli observed,

> It is known with certainty that there is scarcely anything which more greatly excites noble and ingenious spirits to labours which lead to the increase of knowledge than to propose difficult and at the same time useful problems through the solution of which, as by no other means, they may attain to fame and build for themselves eternal monuments among posterity.

Bernoulli accordingly posed the following problem in June 1696. Consider two points A and B on a vertical plane at different heights but not one below the other. (One may imagine two pegs on a wall.) Imagine any curve connecting these two points and consider a ball rolling along the curve from the higher to lower end (say, A to B). Find the curve that will take minimum time to go as previously noted.

Johann gave the deadline of 1 January 1697.

Lest someone argue that the straight line from A to B being the shortest distance would take the least time, Bernoulli informed that it is not the correct answer. However, only one solution was received by the deadline: the solution from the mathematician Leibnitz who had also, like Newton, discovered calculus. To give more time for solutions, Bernoulli extended the deadline.

Isaac Newton had by this time retired from an active role in maths and science. He had taken a civil assignment in London. As the master of the Royal Mint, he was concerned with coinage rather than frontier-level mathematics. Knowing that he may not have seen the advertisement of the problem, Bernoulli sent him a copy by post. Bernoulli may have wanted to humiliate Newton since he knew that Newton having retired may have lost his brilliance.

It is said that Newton saw the problem when he returned from the Mint at around 4 p.m. Although he was tired from a full day's work, he was intrigued by the problem and started working on its solution. He managed to solve it to his satisfaction by 4 a.m. He sent his solution anonymously to Bernoulli. By his revised deadline, Bernoulli had received five solutions. His own solution, then one from Leibnitz, Johann's brother Jacob, Marquis de l'Hospital and the fifth one sent from Britain by an anonymous author. The last solution bore a sign of supreme genius, and Bernoulli after seeing it was chastened and believed to have exclaimed, "I rec-ognize the lion by his paw."

Probability theory

A new branch of mathematics evolved out of the most unlikely source, thanks to the work of Jacob Bernoulli. When a gambler in a casino plays for given stakes, he is not certain of how much he will win or lose. To him, it may appear as if 'win' or 'lose' is quite arbitrary, and there is no control over the outcome. The reality is quite different. Indeed, one can use mathematics to draw certain conclusions. These do not help in forecasting the outcome of a single trial. But one can tell something that applies to a large number of trials. This subject is known as *probability theory*. Jacob Bernoulli wrote a book on the subject that helped considerably in the future development of the field. The book was published in 1713, after Jacob Bernoulli's death.

The so-called Bernoulli numbers form a sequence $\{B_n\}$

$$1, -1/2, 1/6, -1/30, 1/16, -1/30, 5/66, -691/2730, \ldots.$$

It appears arbitrary but is in fact related to probability theory. For every odd n other than 1, $B_n = 0$, while for every odd n other than 0, B_n is negative if n is divisible by 4 and positive otherwise. They were discovered around the same time by Jacob Bernoulli in Switzerland and by Seki Kowa in Japan.

Another fundamental discovery attributed to Jacob Bernoulli is the fundamen-tal number e, which turns up in many situations. For example, if a man borrows 100 Rupees at the interest rate of 10 percent, for one year, he pays back a total sum of 100 (capital borrowed) + interest 10 Rupees. However, if he were to pay by compound interest, he has to face the following limiting process: the interest is computed not yearly or half-yearly but instantaneously – that is, by an infinite number of infinitesimal steps. The result is described by

$$[1+ 10/100\ n]^n = \exp \{1/10\}.$$

Here the answer uses a mathematical method involving limits.

This constant e is a fundamental constant which plays an important part in many parts of pure and applied mathematics.

Here we have given a glimpse of Jacob Bernoulli's contribution. More could be given if space permitted.

Leonhard Euler (1707–1783)

Euler is often regarded as the greatest mathematician after Newton. Having been a pupil of Johann Bernoulli, he was developing into a great scholar of

PHOTO: Euler

mathematics in Switzerland. But realizing that in that country his career would always be dominated by the distinguished Bernoulli family, he left for Russia and later Germany.

Euler's life was marred by many unhappy events including the loss of vision towards the end of life. Despite these handicaps, he contributed to mathematics in several ways. In the year 1775, it was estimated that his pedagogical writings could fill 900 books and in one week he would write one technical article. Because of blindness, he learnt to memorize and work out complicated maths in his mind. He was reputed to have memorized the entire volume of *Virgil* so much so that he could reproduce the first and last line of each page by memory. His versatility was seen from his contributions to geometry, trigonometry, calculus, algebra, arithmetic and number theory, as well as in applied areas like dynamics, theory of light, making maps, music to name some. He is also credited with bringing mathematical writing under a controlled notation so that by following it, others could also benefit. He is also credited with 'the most beautiful equation in mathematics,' which is given below:

$$e^{\pi i} + 1 = 0.$$

It is beautiful because it brings as many as five basic constants of mathematics under one equation $0, 1, i, \pi$ and e.

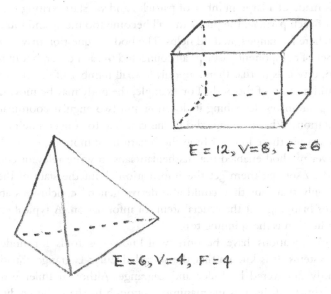

FIGURE 7.2 Euler's rule illustrated with two solids

In Chapter 10, we will describe a problem whose solution by Euler led to new branches of mathematics. Again, Euler was in many ways responsible for the branches of graph theory and topology.

Topology relates to properties that remain valid despite changes of a lot of boundary conditions. For example, the Königsberg bridges (described later) would have continued to give the same answer even if their dimensions were changed. Here is another property of solids that remains the same as again shown by Euler.

Figure 7.2 shows a solid called tetrahedron. It has four vertices, six edges and four surfaces. Denoting these by symbols V, E, S, we find that these are related by the equation V + S = E + 2. What is remarkable is that this equation holds for all solids in the three-dimensional space that we live in. Euler is credited with discovering this relation and giving a general proof of why it is so.

Euler was a respected figure in the court of Catherine the Great of Russia. The anecdote described in the end may not have been a fact, but it certainly fits in with Euler's other achievements!

Joseph-Louis Lagrange (1736–1813)

We now move to another aspect of mathematics. We have already seen that Newton had enunciated three laws of motion. If we imagine a moving particle P, of mass m, with a force F acting on it, our first step would be to set up an 'equation of motion.' In the case of a point particle, there is no difficulty with this step.

However, if we want to describe the motion of a rigid body of finite dimensions, the aforementioned Newtonian method may not be suitable. If we consider the body as made of a large number of particles and we start writing the equation of motion of each particle, the problem will become too messy and intractable.

This is where Lagrange's method helps. The body in question may be made of a large number of component particles all connected to each other, but in reality, the rigidity property tells us that there are only a small number of factors that control the shape and motion of the body. For example, the body may be moving in space with three coordinates describing its location and two angular coordinates telling us its orientation. What Lagrange did in this case was to concentrate on these five items and use only those in describing the equations of motion.

This clever method enabled the mathematicians to write relevant equations of motion and by solving them get the information about the state of the moving body. Not only that, but they could also derive general conclusions about such systems, thus bringing out the crucial items of information. A typical example of such an application is the spinning top.

Lagrange's equations have become well-known as tools for understanding dynamical systems. It is known that the so-called Euler-Lagrange equations were independently discovered by Euler and Lagrange. Although Euler was the first to work in this field, he was magnanimous enough to share the credit with the younger up-and-coming Lagrange.

When Euler was the director of the Prussian Academy of Sciences in Berlin, he recommended along with the support of D'Alembart that the successor to his post should be Lagrange. Following the recommendation, King Frederick invited Lagrange to the post with the following words: *the best king in Europe should have the best mathematician in Europe.*

8
THE TIMELINE OF SCIENCE IN THE NINETEENTH CENTURY

New incentives for science

We have seen how science in the post-Newtonian era grew rapidly. The growth of industrialization was responsible for encouraging science. At the same time in the nineteenth century, the 'classical' topics like astronomy and gravitation no longer occupied the major fields of interest. A timeline of the first half of the nineteenth century can be drawn as follows:

1800 Volta succeeded in making an electric cell (battery), which made it possible to generate an electric current in a controlled and space-saving way.

1805 Dalton initiated physical chemistry starting with his ideas on the atomic structure of matter.

1827 Ohm's experiments led to the notion of resistance to the electric current, a discovery which improved the usefulness of electricity.

1831 Michael Faraday showed how a moving magnetic field produces an electric current, a discovery which opened doors to all kinds of useful applications, starting from domestic to industrial uses.

1848 Lord Kelvin showed the relationship between heat and the internal motion of physical systems, thus laying the foundations of the science of thermodynamics.

1859 Charles Darwin, along with Alfred Russell Wallace, formulated the basic tenet of evolution, which laid a basic framework for biology.

1865–1870 The period when following Kelvin's work, Clausius, Gibbs and others developed the ideas of thermodynamics, including its three basic laws. The concept of the heat engine that emerged from this work was the first step in mechanization.

DOI: 10.4324/9781003203100-9

PHOTO: Faraday

PHOTO: Kelvin

PHOTOS: Charles Darwin, Wallace

PHOTO: Hertz

Electricity and magnetism

Since the first step in 1785 when Charles-Augustin de Coulomb laid the concept of like and unlike charges and the law of repulsion (attraction) between like (unlike) electric charges, the finding that these forces are following inverse square law seemed to indicate that the electric forces may be similar to gravitation. Subsequent experiments showed that this was not so. Indeed in 1864, James Clerk Maxwell showed that electricity and magnetism can be unified in a symmetrical set of equations and the outcome of the combination was the important conclusion that the electromagnetic disturbances travel together as an *electromagnetic wave* with the speed of light.

This prediction was experimentally tested by Rudolf Hertz in 1887 when he could produce electromagnetic waves in the laboratory. While it was a great scientific achievement, it was a precursor to an even greater one – namely, the theory of relativity, which we shall describe in Chapter 13.

The electromagnetic wave

The studies of electricity and magnetism had progressed during the eighteenth and nineteenth centuries on more or less parallel lines. Thus the various results made one suspect that there is a deeper connection between electric and magnetic fields. Michael Faraday's experiments demonstrated that if magnetic flux through a closed electric circuit is changing, it produces current through the circuit. In short, the magnetic field appeared to have a connection with electricity.

PHOTO: Maxwell

These ideas were followed by James Clerk Maxwell in the 1860s. Writing all the relationships connecting electricity and magnetism, Maxwell had arrived at a nearly symmetric set of equations. Writing these equations in a vacuum gave a form that suggested adding an extra term. In fact, a look at these equations convinced Maxwell that adding an extra term will make the symmetry perfect.

Thus, he predicted that a *displacement current* also plays a part in electromagnetic interactions. Moreover, these equations lead to the conclusion that both electric and magnetic fields travel as a wave, which may be called the *electromagnetic wave*. And a fundamental conclusion emerged that the wave travels with the speed of light. Hertz actually created a laboratory demonstration of how such a wave can be prepared and transmitted.

The nature of heat

While following Newton, scientists began to appreciate the role of dynamical energy. Suppose you are given a fragile glass mirror and a small metallic ball. You are asked to hit the mirror with the ball faster and faster. There will come a stage when the ball hitting the mirror cracks it. One can say that the mirror cracked when hit by a sufficiently energetic ball.

Newton's laws of motion helped clarify the role of energy of motion. It is defined by the formula

$$T = \frac{1}{2} M v^2,$$

where M is the mass of the ball and v its velocity. Because it is linked with motion, it is called *kinetic* energy.

There can be other forms of energy. For example, Newton's laws of motion describing a ball tossed up in the air bring into play *two* kinds of energy. For, we have the governing equation as

$$\frac{1}{2} M v^2 - GMM_E / R_E = - GMM_E / R_0,$$

where the second term on the left-hand side is the *gravitational* energy of the ball arising from the gravitational force of attraction of Earth on the ball. The negative sign before this term indicates that the gravitational force is one of attraction. Thus, we have another example of energy, *called gravitational potential energy.*

When we witness a waterfall, for example, we have water drops moving slowly at the upper end and much faster-moving drops at the bottom end. This is an example of the conversion of one form of energy (gravitational) into another (kinetic). The underlying rule that emerges from such examples is the law of *conservation of energy* (Figure 8.1).

FIGURE 8.1 The Bhakra Nangal Hydroelectric Dam. Water falling down drives underground turbines which produce electricity. Thus, we have gravitational energy converting to hydro to electrical energy.

We have come across other examples too, like the motion of electric charges where the electrical energy between like charges is positive and that between opposite charges is negative.

In the eighteenth century, physicists began to formulate rules for another form of energy which we may call *heat* energy. When James Watt discovered that the lid of a vessel in which water was being heated was tossed around, the energy supplied to the lid came from the heat supplied to the water in the vessel.

The case of heat energy, however, has certain restrictions, and these are expressed through the *laws of thermodynamics*, which we describe next.

The Zeroth Law: This law essentially defines what we mean by temperature. Imagine we have a container filled with gas. The gas contains randomly moving atoms and molecules. If we heat the gas, the internal motion becomes more agitated. This is expressed by saying that the temperature of the gas has risen. Thus, we can ascribe a temperature to the gas provided it has settled down to a state of equilibrium.

Here equilibrium means that each constituent particle of the gas keeps changing its state through collisions but the overall profile of all the particles remains the same.

The First Law: This law tells us that heat energy is part of the different sources of energy (dynamical, gravitational, etc.) and is subject to the law of conservation of energy. For example, in an engine, the heat energy is converted to energy of motion.

The Second Law: This law imposes a unidirectionality on how heat is exchanged with other energies. The golden rule is that the change in the system will always be towards greater disorder. The parameter that measures order and disorder is the *Entropy* of the system. Thus we can express the second law by saying that any change in the system will be towards keeping the entropy the same as before or increasing it.

The Third Law: It tells us that no way can we manage to bring the temperature of the system down to value zero. Thus, this is one of the 'impossibility' theorems.

We will make a few comments on these laws.

First. We are provided with a natural scale of temperature. Called the *absolute scale*, it is related to the Celsius scale as follows:

0^0 Celsius $= 273$ Absolute,
-273^0 Celsius $= 0$ Absolute.

The third law then tells us that it is impossible to reach the temperature of -273K. The letter K stands for Kelvin. Lord Kelvin was one of the pioneers of the science of thermodynamics.

The fact that one can use a source of high temperature to produce motion led to its application to drive trains. The first law of thermodynamics tells us that heat can be converted to motion – the engine is the place where this conversion is carried out. But the second law tells us that there are also certain ifs and buts attached to it! Thus, it can be shown that the temperature of the heat source in the engine must be greater than the temperature of the surroundings to make it possible for conversion to take place. Indeed, if the temperature is allowed to rise so that it reaches that of the source in the engine, the energy-generating process will come to a halt.

Thus, thermodynamics prompts the research to proceed in a way that the ratio of the temperatures of the sink and the source has as low a ratio as possible. Here the source produces heat while the sink receives it.

Avogadro's number

This is an example of the wide range of applications of thermodynamics. In chemistry, when one is looking at gases of different masses, one notices the following result. Imagine that the molecular weight of a gas is W. That means that a molecule of the gas weighs W times the weight of a hydrogen atom. The quantity W gm of this gas will contain N molecules of the gas. Thus, 32 gm of oxygen will contain N molecules of gas. Using thermodynamics, one can show that this number is the same for all gases. This result was first deduced by Avogadro, and the number is called *Avogadro's number*. This recognition made a lot of simplification in understanding and quantifying physical chemistry. The Avogadro number is 6.0221415×10^{23}. This also tells us that the mass of the hydrogen atom in grams is the reciprocal of the Avogadro number.

We now consider a pioneering achievement in the field of biology.

The science of genetics

When a small baby is introduced to a family friend, the doting parents and the visitor spend some time speculating whom the baby resembles. The assumption, well backed by observations, is that there are features of the baby which show points of resemblance with parents, grandparents and other close relations. In fact, such observations are very common and have been around for hundreds of years. Apart from humans, one can see them in various fauna and flora, resulting in the expectation that there may be some scientific law governing them.

A beginning was made to understand how physical characteristics are transmitted from one generation to another by Gregor Mendel. In his lifetime (1822–1884), he did not receive credit for his ideas and supporting evidence but today he is a big name, and his ideas have blossomed out into the subject of *genetics*. Indeed, any account of science in the nineteenth century would be incomplete without the inclusion of genetics. Here we will broadly describe the early period and the pioneering role of Mendel.

Born in Austria of farmer parents, Mendel had aimed for a religious career. He had studied mathematics and science at the University of Vienna. He had also worked as a gardener. Both these qualifications helped him in his later work. He was a good and dedicated teacher and a scientist, as well as a priest. Technically speaking, he was an Augustinian monk in the Czech Republic, and in the seven-year period starting in the year 1856, Mendel conducted his experiments. Despite their success, they remained unknown and were rediscovered some 50 years later by three scientists named Erich Tschermak, Hugo de Vries and Carl Correns. What did the experiments involve?

Experiments with plants

Mendel was interested in a special type of plant. The so-called hybrid plant, which is typically obtained by combining two different plants into one. He chose for his study typically "pea plants" because of their property of growing fast. He acted as a pollinator, making the new generation plant by careful controlling. He worked with a large number of plants so that he could observe their properties under different control conditions. Thus, he could observe seven different properties in the variety of plants he was experimenting with.

These were

(1) colour of the flower (purple or white),
(2) position of the flower (axial or terminal),
(3) stem length (short or tall),
(4) seed shape (round or wrinkled),
(5) seed colour (yellow or green),
(6) pod shape (constricted or inflated) and
(7) pod colour (yellow or green).

He cross-pollinated different species, with some having the same characteristics while in some cases the characteristics would be different. He observed and recorded the types that came from the union, thus collecting vast data on combinations. Based on his observations, Mendel proposed a theory in which inheritance from a parent played a part. Thus one offspring receives one flower colour from each parent. Mendel used the word *allele* for the individual trait from one parent. The alleles all taken together form a *gene*. Mendel observed that some traits are more common than others. Also, some tend to dominate others and will show up if present, while some tend to lie hidden. He called them *dominant* and *recessive*, respectively.

Mendel had a lot of data and statistics (compared to earlier work) and so had his deductions in the form of two laws: *the law of segregation* and *the law of independent assortment*. The former showed that dominant and recessive traits are passed on from parent to offspring *randomly*. And the latter showed that traits were passed on independently. In any case, Mendel's findings differed from earlier (not so systematic) work, even though he did not do anything to promote his own ideas. Indeed, this was perhaps the reason that he did not get credit for his research till well after his death.

Mendel's observations

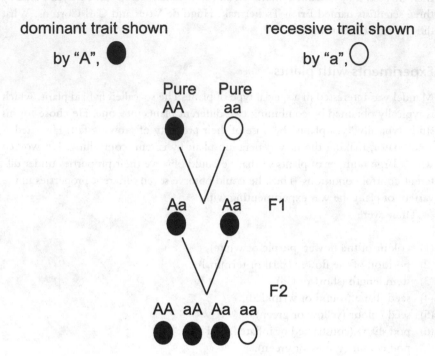

dominant trait shown by "A",

recessive trait shown by "a",

Pure AA

Pure aa

Aa   Aa   F1

F2

AA aA Aa aa

FIGURE 8.2   Mendel's experiment with different generations of plants.

Mendel's ideas on heredity

Gregor Mendel carried out a large number of studies on pea plants, and based on those observations, arrived at a basic set of rules about hereditary traits. He used the garden of his monastery to conduct these experiments, which basically involved selective crossbreeding. Before him, the general belief had been that parents provide 'essences' which were mixed in the 'offspring' just like the blending of paints blue and yellow producing green. Mendel's idea was that heredity is based on discrete units of inheritance so that in any individual, these units called 'genes' act independently of one another. What will happen in deciding the inheritance of a particular trait is how these genes are passed on from parents to the individual. For any trait, the individual inherits one gene from each parent. These units in their alternate combination are known as 'alleles.' If the two alleles for a given trait are the same, then the person is called 'homozygous' for that trait, whereas an individual with two different genes is called a 'heterozygous.'

PHOTO: Mendel

Mendel selectively cross-bred the pea plants with chosen traits for several generations. (The reason for choosing pea plants was their short lifetime for a generation.) The traits looked for included stems short or long, round peas or wrinkled peas, white or purple flowers, etc. He saw that the next filial generation called F1 was made of individuals of one trait only. But when this generation is interbred, its offspring, called F2 by Mendel, showed a 3:1 ratio regarding any such trait. That is one group would have the trait of one parent while the other will have the trait of the other parent, and these groups will have the aforementioned ratio.

Mendel assumed that genes can be of three possible pairings: AA, Aa, aa of which A is the dominant and a is recessive. That is, A occurs more often in the population than a. The recessive trait will result if both 'factors' are recessive. Mendel had defined the factors in this context.

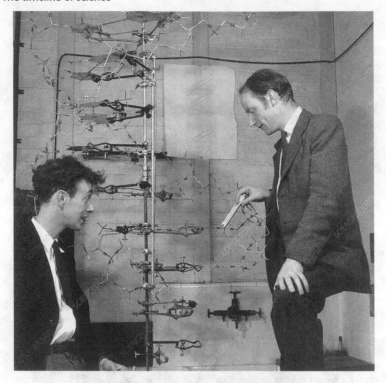

PHOTOS: Watson and Crick, Rosalind Franklin

Further experimentation

Genetic experimentation like Mendel's inspired R. A. Fisher at Cambridge not only in further such studies but also in the application of statistical methods. Born in 1890, Fisher was keen to do wartime service but was disqualified on medical grounds (weak eyesight). While he reviewed books to earn a livelihood, reading them got him interested in the use of probability theory in the interpretation of genetic data. His methods met with a lot of success, which made statistical techniques very popular for testing hypotheses. Every human being has 46 genes of which half are derived from their mother and half from their father. Which 23 out of the parental 46 genes are so derived is not fixed, which explains that even two brothers with the same parents may have differences.

Mendel's article on his main work appeared in 1866, the same year in which T. H. Morgan, another would-be genetic scientist, was born. Morgan worked in distinguished places like Cal Tech and the universities of Johns Hopkins and Columbia. He used the drosophila, a kind of fruit fly for many of his studies. This fly is short-lived and therefore convenient for such experiments.

The passage of genes from parents to the offspring happens randomly. For example, if of the parents one has grey eyes while the other has black or dark eyes, then what colour eyes will the offspring acquire? Half probability comes from the grey eyes source while half comes from dark-grey or black eyes. Because the dark-grey colour happens to have a dominant feature, the offspring will have dark-grey eyes. But the other (grey) eye feature remains recessive as the person of the next generation is formed. That person on mating with another similar gene structure will have a gene colour grey or dark grey. The probability of their child having grey eyes is then one-fourth. Thus, it is possible for a child born of parents with dark-grey eyes to have grey eyes. The science of combinatorics plays a useful role here.

The episode of Queen Victoria and her descendants is an example of genetic effects. The illness known as haemophilia was found in several of her children and grandchildren. A gene of Type X is responsible for it. However, it is known that an XX pair is found in females and XY in males. It is known that females do not suffer from this disease but they can be carriers. Victoria's daughter who became Queen of Prussia was a carrier, but her two sons did suffer from it. Her second daughter who became queen of Russia was a carrier, and her son was a victim of the disease. Other examples from this royal family can be given. This shows how hidden genes can cause trouble.

The discovery of DNA structure

The progress on the genetic front was the indicator of the progress in biology. A major development in this field was the discovery of deoxyribonucleic acid (or DNA in short!) in 1869. The early days in molecular biology were concerned with the question as to whether the genes were made of proteins or nucleic acids. The final answer took a long time, about eight decades, to find out. Francis Crick and

James Watson are credited with the unravelling of the structure of DNA in 1953. They along with Maurice Wilkins were awarded the Nobel Prize for this discovery. Another scientist who should have shared the prize but missed out because she died early (in 1962) was Rosalind Franklin.

Structure of DNA

As mentioned earlier, the structure of DNA (and RNA) took some time to establish. DNA was known to be a long polymer composed of four subunits that

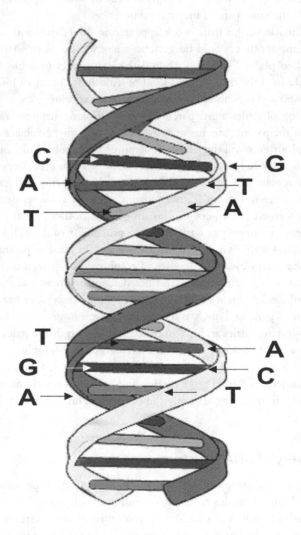

FIGURE 8.3   A schematic figure showing the nature of the typical DNA molecule.

resemble one another chemically. A better understanding of the DNA structure came with X-ray diffraction analysis early in the 1950s. This method was able to determine the three-dimensional atomic structure of the molecule. Early work showed it to be two polymer strands wound into a helix, as shown in Figure 8.3. The DNA molecule has two long chains of polynucleotides consisting of four nucleotide subunits. Each of these chains is called a *DNA chain* or a *DNA strand*.

Nucleotides are organic molecules used for making nucleic acid polymers DNA (deoxyribo-nucleic-acid) and RNA (ribo-nucleic-acid). These biomolecules are part of all living forms. Nucleotides are the building blocks of nucleic acids and are composed of three subunits:

a nitrogenous base (nucleobase) + a five–carbon sugar (ribose/ deoxyribose) + one from phosphates.

The nucleotides are connected in a chain through the sugar and phosphate alternately like beads on a necklace. There can be four bases called *adenine (A)*, *cytosine (C), guanine (G)* and *thymine (T)*.

The way the bases are linked results in a chemical polarity of a typical DNA strand.

It is usual to look at the links as a cylinder with a protruding knob at one end and a matching hole at the other so that in a linkage the knob of one cylinder goes into the hole of another. The knob is usually called a 5' phosphate while the hole is called 3' hydroxyl. Thus a DNA chain ends in one of these two and often one refers to the end by 3' end or 5' end.

Like the rungs of a ladder, one can think of two chains held together by covalent hydrogen bonding. Thus the sugar-phosphate bonding (called backbone) is on the outside of the helix while the bases are inside being linked by the hydrogen bond. This restricts linkages to A with T and C with G. One can look upon the whole arrangement as energetically the most favourable (corresponding to the criteria of least energy in dynamics!). To maximize the efficiency, the two sugar-phosphate backbones form the double helix by winding around with one complete turn every ten pairs.

9

ON THE COSMIC FRONT

Geocentric or heliocentric?

In the intellectual debates which were inspired by the arguments presented by Nicholas Copernicus back in the fifteenth and sixteenth centuries, the crucial point was whether Earth as a planet was at rest and the other planets, as well as the Sun, moved around it, or it was the other way around and all planets, including Earth, orbited around the Sun, which was the dominating body of the planetary system and was at rest.

The view popular with the intellectuals of the time was the former and it had received the sanction of the religious authority in Europe. So much so that it was considered sacrilege to hold any other view. History tells us that Copernicus's book *Revolutionibus Orbium Coelestium* held the contrary view and was banned by the pope. Any support to it in public or private was, to put it mildly, frowned upon.

Amongst the small minority who subscribed to the Copernican viewpoint was the seventeenth-century Italian scientist Galileo Galilei. To appreciate the qualification "scientist" we remind ourselves that science as a subject evolved from man's attempts to understand how nature functions and that its basis is experimentation supported by theory. Thus theory must make a prediction about how nature would behave under prescribed conditions. The experiment is then expected to test the validity of that prediction. In this procedure, there is no room for any religious authority dictating its own rules.

Galileo was probably the first major scientist to appreciate the role of experiments. His book with the translated title *The Dialogue Between Two World Systems* shows clearly how the world model based on the traditional view that had come from the teachings of the great Greek philosopher Aristotle was wrong. Instead, it argues for different sets of beliefs. The important thing is that the alternative

DOI: 10.4324/9781003203100-10

proposed by Galileo is supported by experimental evidence, which the traditional model is not.

In the prevailing atmosphere of seventeenth-century Europe, this conclusion was regarded as against the teaching of religion, and Galileo was subjected to rigorous questioning by the Inquisition. In the end, he "recanted"; that is, he publicly admitted that Earth is stationary and that he had been wrong and the Church had been right on this issue. That way he was spared the terrible tortures that were normally applied to those who were guilty. Instead, he had an enforced monastic life sentence of living in his own house with no outlets for his ideas.

In retrospect, we can ask what evidence Galileo had to support his view that Earth is not fixed in space but moves around the Sun. Indeed, the Inquisition had asked him the same question. This was to be expected since the whole issue of Galileo's trial revolved around this point. When Galileo answered this question, he actually did not possess the correct answer.

Galileo's reply was somewhat like this. If we walk holding a glass full of water, the disturbance caused by walking would make the water in the glass shake a little, and if we are not careful, water will spill out. The fact that tides make ocean water jump up and down indicates that Earth carrying the water is not stationary. Q.E.D.!

This reply is not correct since the tides on Earth are now known to be caused by the gravitational attraction of the Moon and the Sun. The wrongness of Galileo's explanation is shown by the timings of tides which could not be predicted by Galileo. Two effects show in a direct and factually correct way that Earth moves: (1) aberration and (2) parallax.

As a postscript, we may mention that a review of Galileo's trial by Inquisition was conducted by the order of the pope in 1979. The experts on the review panel exonerated Galileo and concluded that the decision of the Inquisition was not justified. The panel of course noted the wrong answer given by Galileo but despite that his assertion about the motion of Earth was correct. It is, therefore, appropriate to agree with the sentence Galileo is said to have uttered in a whisper about the motion of Earth: *Eppur si mouves* (But it moves).

Aberration

There are two ways in which today's astronomer can check whether Earth moves in relation to the stellar background. The first is called aberration. To understand how it works, imagine the situation where a man is driving a car in rain which is falling vertically down. To the driver of the car, however, the rain falls in a slanted direction, as shown in Figure 9.1. This is because the car's velocity is vectorially subtracted from the velocity of the rain. For a stationary car, however, the rain obviously falls vertically down. Also, if the driver reverses the direction in which he is moving, the rain will also appear to him to be coming from a different slanted direction. We now apply this rule to Earth moving against the stellar background.

If Earth was stationary, the relative speeds of stars today and six months later would be no different. In the case of a moving Earth, however, the direction of

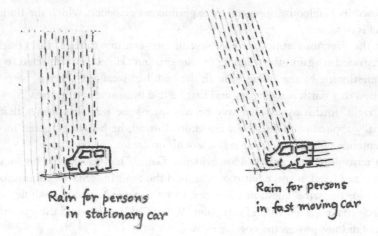

Rain for persons
in stationary car

Rain for persons
in fast moving car

FIGURE 9.1   Rain falling vertically down, an example of aberration. The first image is
of the rain as seen by the driver of the stationary car. If the car moves fast in
a certain direction, the rain appears slanted to the driver. Use this analogy
for light coming from a star to the moving Earth. The observed minute
effect is called aberration.

starlight should change, as in the case of the car driven in rain, since Earth's direction of motion today and six months later would be opposite. The light coming
from a star will likewise change its direction for an astronomer observing from the
moving Earth.

This effect is of course very small but it became measurable in the eighteenth century when accurate astronomical instruments needed for such observations became available. In 1725, James Bradley carried out the measurements
of the star *Gamma Draconis* and the change in six months could be detected.
This was the first direct evidence for the motion of Earth against the stellar
background.

Parallax

The second method of proving that Earth moves was the one Aristarchus had
attempted nearly two millennia ago. This method is now called the parallax
method, and its principle is as illustrated in Figure 9.2. The figure shows a person
driving past a monument located about half a kilometre away. The mountains in
the background serve as reference points for the person driving past. Thus the
motorist sees the line of sight to the monument O slowly rotate from AO to BO
as he moves from location A to location B. The background of distant mountains
helps him to see the effect, which is known as parallax. The angle of rotation AOB
measures the magnitude of the effect.

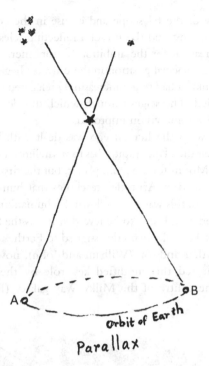

Parallax

FIGURE 9.2 Parallax observed. A star observed from Earth on nights separated by six months should show a slight change of direction if Earth is moving. This effect is called parallax.

Now imagine an observer looks at a specific nearby star O at two instants separated by six months during which period Earth moves from A to B. The angle AOB gives the magnitude of the same parallax effect. The direction of the star can be measured with respect to the stellar background of faraway stars. Such a background will not change as Earth moves. The change, in any case, would be small: whereas angle AOB will be small but measurable. The measurement of the first such parallax was by Friedrich Bessel of star *61 Cygni* in 1838.

Note that this was precisely the method recommended by Aristarchus. It failed because he had underestimated the distance of a star from Earth. As a result, he expected a much larger value for the angle AOB. The measuring instruments in his time were not very accurate, and so the observers reported a negative result.

Astronomy with telescopes

In 1609, Galileo first used the telescope to view the cosmos. He realized the potential of that instrument and soon made discoveries that were beyond the capacity of naked-eye astronomy.

Galileo's discovery of the telescope and its use in the cosmos was the beginning of modern astronomy, and the use of a reflecting telescope by Newton was the opening of the heavens for the ambitious astronomer. However, man's belief that he was occupying a special position in the universe began to get shattered. All along, though, humankind had to get increasingly relegated to an ordinary position as the cosmos unfolded. The stages through which this demotion occurred came about as humankind's cosmic vision improved.

The very first blow to the human ego was dealt with by Galileo's telescope when it confirmed that the planet Jupiter has four satellites. Later observations were to reveal many more Moons for the giant planet, but the first discovery had already done the damage. Trained by Aristotle's teachings that humankind on Earth is at the centre of the cosmos, this was a comedown. So humankind, reluctantly perhaps but accepting a revised role, learnt to believe that it was the Sun and the planetary system that played the key role formerly assigned to Earth.

The Herschels, father and son (William and John), now supported by much larger telescopes boosted this modified key role by their claimed discovery that the Sun is at the centre of the Milky Way galaxy (Figure 9.3 a, b). The

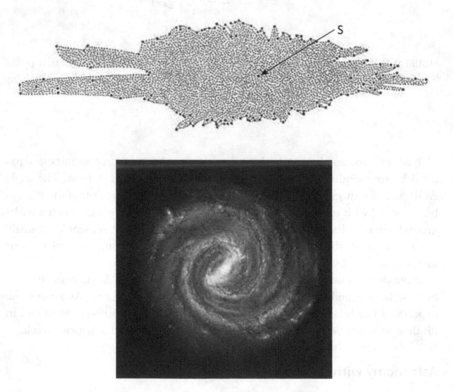

FIGURE 9.3 a, b: Herschel's map of the Milky Way. It shows the Sun close to the centre. With better and wider observation today, we have the adjoining picture of the galaxy with the Sun about two-thirds of the way to the boundary.

promotion to the centre position of the galaxy may have restored some of the lost ego but even that was not to last long. The cosmic distances are somewhat tricky to estimate, and by the beginning of the twentieth century, the revised distances placed our Sun not at the galactic centre but nearly 30,000 light years from it and nearly two-thirds of the distance to the boundary. (A *light year* is the distance travelled by light in one year, which works out to be around 10,000,000,000,000 kilometres.)

But as humankind nursed another blow to their claimed role of being a special inhabitant of the universe, a different role was being touted for them. The most sensitive telescopes were showing some faint nebulous cloudlike patches superposed on the location of the Milky Way. The general claim was that these 'nebulae' were made of gas clouds or star groups, all part of the Milky Way. Thus, the claim generally subscribed to was that all that we see through our telescopes belong to our galaxy.

However, a small minority was of a different view. It felt that some of these nebulae were outside our galaxy, being very far away and hence so faint. Again a controversy erupted: are these nebulae all part of our galaxy or are some of them galaxies in their own right looking so faint because of being far away? The majority view is well stated in the popular book of Agnes Clerke:

> The question whether nebulae are external galaxies hardly any longer needs discussion. No competent thinker, with the whole of available evidence before him, can now, it is safe to say, maintain any single nebula to be a star system of coordinate rank of the Milky Way. A practical certainty has been attained that the entire contents, stellar and nebular of the sphere belong to one mighty aggregation, and stand in ordered mutual relations, within the limits of one all embracing scheme.
>
> From The System of the Stars 1905

This view so categorically expressed did not last more than two decades. By the 1920s, the existence of external galaxies became established, and the entire picture was due for a major overhaul.

We will next look at astrophysics of smaller structures like galaxies, quasars, stars, intergalactic and interstellar dust.

Stellar structure

The best success story of astrophysics is of the internal structure of stars. It started in the nineteenth century with the question, "What makes the Sun shine?"

The observations show the Sun's luminosity to be 4.10^{33}erg per second. The age of the Sun is estimated at 5.10^9 years; that is around 1.5×10^{17} seconds. With the aforementioned rate of emitting energy so far, the Sun has used up energy to the tune of

$$1.5 \times 10^{17} \times 4 \times 10^{33} = 6 \times 10^{50} \text{erg}.$$

If the Sun is to go on shining for an equal length of time, it must have an energy provision of twice the above value. How does the Sun manage this?

Two leading scientists, Lord Kelvin in Britain and Baron von Helmholtz in Germany answered this question by pointing to the Sun's gravitational energy reservoir. A body of mass M and radius R has gravitational energy of the order of $-GM^2/R$. The Kelvin–Helmholtz contraction hypothesis had the Sun slowly shrinking from a dispersed state, releasing gravitational energy. Notice that because of the negative sign, as the body shrinks, its energy storage is reduced.

For the Sun, the mass is 2×10^{33}g and the radius 7×10^{10}cm, giving a total energy reservoir of -4×10^{48} erg. What does it all mean? The previous figures contain the answer. If the Sun were to call upon its gravitational energy reservoir to account for its radiating power, it would last for less than a percent of its present age. Thus, the gravitational hypothesis does not work.

The failure of the Kelvin–Helmholtz contraction hypothesis led to a rethink of the stellar model. Eventually, Eddington in the 1920s came up with a working model. In this model, the Sun is a ball of hot gas, ideally a form of plasma which is hydrostatically balanced so as to have a temperature of the order of ten million in the central region but down to a few thousand at the outer boundary.

This model led Eddington to propose that the energy stored in the centre of the star is released through nuclear fusion reaction converting four nuclei of hydrogen

PHOTO: Kelvin

PHOTO: Helmholtz

PHOTO: Eddington

to one nucleus of helium. The energy released this way is quite adequate to keep the Sun shining at its present rate for a total of 12 billion years. As mentioned earlier, the present age of the Sun is estimated at 5 billion years.

Origin of elements

The nuclear reaction referred to previously gives a clue to the answer to another question: how and when were different atomic nuclei made? There were two possibilities. While discussing the early universe, we shall see that the age range up to three minutes provided the conditions suitable for the production of light nuclei. Provided certain physical conditions were met, the origin of deuterium (the name given to 'heavy hydrogen'; Figure 9.5) and a few light nuclei could be explained this way. We will return to this scenario later when discussing the nature of dark matter.

The triple-alpha process

The way Fred Hoyle arrived at the previous result is itself of interest. The presence of life in the universe, as indicated by life on Earth, suggests that carbon is one of the important constituents thereof, and so we have to find a cosmic process for

PHOTO: Hoyle

producing carbon. At first, it was thought that three helium nuclei would naturally combine to form carbon. But there was a hitch! In a collection of randomly moving alpha particles (this is a name given to the helium nuclei), the chance of three particles arriving at the same spot at the same time was very small, thereby making the process not of much use. Hoyle's way out of the problem was to propose that the three-body collision can work *provided the process resonates*.

Resonance comes in different forms in our daily life. A musical instrument resonates when its natural frequency of vibration matches the externally imposed frequency. In that case, the instrument responds in an amplified fashion. If a child sitting on a swing is given periodic pushes by a parent say, and if the period of pushing matches the natural frequency of the swing, the amplitude of the swing greatly increases. Likewise, in a resonant nuclear reaction, the equality of the energy of input and output can produce resonance, which leads to the reaction going fast. So Hoyle reasoned that if the combined energy of three alpha particles matches the energy of carbon produced, the production will be fast and would compensate for the rarity of three-particle collision. For this to work, there should exist an excited state of carbon nucleus with energy equal to the total energy of the three alpha particles.

Armed with this argument, Hoyle approached experimental nuclear physicists and asked them to look for an excited state of carbon with extra energy implied by the resonance argument. Somewhat bemused by this unexpected prediction, the nuclear physicists finally took the trouble of setting up the necessary experiment. Ward Whaling and Willy Fowler at Caltech finally succeeded in finding the predicted state. To Hoyle, it was a demonstration of an anthropic argument – that is, an argument leading to a prediction deduced on the assumption that human beings exist in the universe!!

The purpose of the previous exercise was to demonstrate that cosmic processes are capable of producing all the chemical elements that go into making a human body.

Following Fred Hoyle's brilliant deduction, the further details leading to the formation of chemical elements were worked out by four scientists working together, viz. Margaret Burbidge, Geoffrey Burbidge, William Fowler and Fred Hoyle. This important paper or its authors are often referred to as B^2FH.

The evolution of stars

The work of B^2FH, which we briefly referred to earlier, shows that stars serve as thermonuclear reactors for making elements. As their internal composition changes, the stars also change their physical properties. Thus as the Sun-like star converts the hydrogen in its core to helium, it grows in size, becoming a 'giant' star. This reduces its surface temperature, making it appear more reddish. Thus the so-called *red giant* stars observed by astronomers can be understood as part of this evolutionary scheme. It is expected that the Sun will become a giant and will grow

PHOTO: Burbidge, Burbidge, Fowler, Hoyle together

so big that it will swallow the inner planets Mercury, Venus, and Earth, possibly Mars too.

The more massive stars would become even larger. However, there is a bifurcation of the future at this stage. Stars like the Sun, and even as massive as five to ten times the Sun, end up as *white dwarfs* when they have exhausted all nuclear sources for generating energy. Unable to hold their original sizes, their self-gravity shrinks them until their size is as small as a percent of the size of the Sun.

The second route is followed by the more massive stars, wherein the huge size of the star makes it unstable and it explodes. Such a star is called a *supernova*. Figure 9.4 featuring the Crab Nebula describes a spectacular supernova and the stories associated with it.

An exploding star leaves behind a dense core, mainly of neutrons, and in general, it is expected that the aftermath of the supernova will be a *neutron star*, which is a highly dense (typical density being *a million billion* times the density of water) star which is seen more as a pulsating radio star, called *pulsar*.

Can such stellar explosions be seen? Indeed, if we want to see such a spectacular event, we will need it to be close to us, say a few hundred light years. But a very close event, say less than 30 light years, has to be avoided. Cosmic rays coming out of such an explosion can be dangerous to life if the explosion is too near! But even from a distance, the event can be worth observing.

An explosion in space

The night of 4 July 1054 was to contain an eventful occasion so far as the Chinese observers were concerned. The professional ones were appointed by the emperor of China's Sung Dynasty, with a specific purpose: in case they saw something unusual happening in the heavens, they were to record it and report immediately to the emperor. Why?

Because the common belief was that if the emperor took a wrong step and broke the code of conduct, he would receive a warning sign from the heavens. This enabled the emperor to take corrective action or atone suitably for his action. Of course, if he missed the sign somehow, his lack of action would be interpreted as disrespect to the 'Higher Powers above' and that would spell badly for the emperor.

So the emperor had arranged to maintain a steady watch on the heavens so as not to miss any sign from above.

The sign came in the form of a 'visit by a guest star.' There was a sudden brightening in a specific direction as if a new star appeared there. The brightening went on increasing for a few nights before it started diminishing until the 'star' disappeared from the observer's sight. (We need to remind ourselves that those were the pretelescope nights – telescopes came into use much later, in the year 1609.) As mentioned in the article titled "History of the Sung Dynasty" by Ho Peng Yok in the 1962 issue of *Vistas in Astronomy*, 5, 184," the record says, "On a Chi-Chhou Day in the fifth month of the first year of Chi-Ho reign period a 'guest star' appeared at the south east of Thien-Kaun measuring several inches. After more than a year it faded away." One wonders how the emperor interpreted this event.

The unexpected appearance and later disappearance from the sky of this object likened it to a guest star. Like a guest, the star came and went. But of course, a look at the phenomenon from a modern standpoint tells us that it describes the supernova phenomenon. A star, especially at least five to ten times as massive as the Sun eventually explodes, at which time its brightness is extremely high, even exceeding that of a galaxy of a hundred million stars! The star that was seen by the Chinese was no doubt a red giant star with low luminosity but which grew in brightness as it exploded. While the aftermath of the explosion has provided astronomers considerable food for thought, some of us have also wondered as to whether other observers in different parts of Earth did see and record the event.

Two independent sources have indicated that the aforementioned event might have been seen in two different parts of the world. One is in the Middle East and the other on the American continent.

The latter sighting was reported in 1955 by William C. Miller, a research photographer in the Mount Wilson and Palomar Observatory. He presented evidence that the Pueblo Indians in the south-west United States had seen the event and recorded their impression on the stones nearby. One is a pictograph, an image made on rock with paint or rock piece that writes like chalk. The other is a petroglyph – i.e., an image chiselled on rock with a sharp implement. The pictures

FIGURE 9.4 The Crab Nebula. Observed in modern times. This is believed to be the remnant of the exploding star observed by the Chinese on 4 July 1054.

convey the impression of the observers of the event. In this case, we see a lunar crescent with a round object nearby. The crescent corresponds with the lunar shape on the night when the star exploded. The round object was interpreted by Miller as the 'guest star.' Although primitive in nature, the evidence is indeed suggestive.

The second line of evidence was brought to this event in 1978 by Elinor and Alfred Lieber from Israel and Kenneth Brecher from the Massachusetts Institute of Technology in the United States. These authors drew attention to the writings of Ibn Butan, a Christian physician from Bagdad. As he was interested in the hypothesis that many illnesses on Earth are related to cosmic events, he kept good records of both the cause and effect. His findings led to the apparent correlation of the suddenly brightening star and the epidemics on Earth. As per his notes, the star is none other than the supernova found by the Chinese.

What about Indian sightings? The date of the supernova shows that India was enjoying prosperity in astronomy, and so an event like this should not have gone unnoticed. It could be argued that with July being a peak monsoon period in India, because of cloudy skies, astronomers there may have missed the chance to see the event. However, it is unlikely that absolutely no observations were possible. Even in a Monsoon period, some parts of the sky can be cloud-free. Remembering that this event lasted for several weeks for naked-eye astronomy, surely there must have been some mention of it somewhere. Two of us (myself and Saroja Bhate, a Sanskrit scholar) in 2001 carried out an extensive search of possible references to supernovae in old writings but to no avail. Although we looked for mainly Sanskrit records, it is quite likely that information was recorded in a local language. Another view is that the information was given in local metaphors as suggested by librarian Shylaja and her colleagues from Bangalore.

The expanding universe

While astronomers were concerned with measurements of distances of ages and sizes of galaxies in order to estimate the overall size of the universe, a different line of measurement began to assert itself: the spectrum of a typical galaxy showed that the spectral lines were *redshifted*. That is, a given spectral line may show a wavelength *longer* than its local laboratory value. Thus if the observed wavelength was λ and the expected laboratory wavelength was λ_0, then the quantity z defined by

$$1 + z = \lambda/\lambda_0$$

turns out to be positive and is called the redshift. Barring a few nearby galaxies, the rest showed this phenomenon. Also, z appeared to increase with the distance of the galaxy.

What did this mean? If we make a simple deduction based on the well-known phenomenon called the Doppler effect (see the following section), we conclude that all redshifted galaxies are receding from us. Moreover, the farther the galaxy is from us, the larger its speed of recession. Such a conclusion led astronomers to the concept of the expanding universe.

Doppler effect

When we stand on a railway platform and see an express train whiz past, the engine of the passing train whistles, and the whistle pitch appears to us stationary observers on the platform to drop in shrillness as the engine passes us. This is known as the Doppler effect in sound. To understand it crudely, imagine the whistle sound being communicated by waves of alternate compression and rarefaction. So long as the engine is approaching us, the waves between us and the engine get more compressed with the result that the whistle sounds shriller. Similarly, the opposite happens as the engine recedes from us.

A similar phenomenon happens as we receive light waves from a moving source. The correspondence between light and sound is understood by the shrillness of sound and shift towards the blueness of the spectrum. Likewise, the flatness of sound would correspond to a shift of spectrum towards the red end. The extent of shift would tell us the recession speed of the light source.

Most galaxies show the spectrum shifted towards the red end. The engine analogy thus implies that all these galaxies are moving away from us.

We shall discuss (see Chapter 13) the special and general theories of relativity proposed by Einstein. Einstein, as we shall see, believed in the concept of a static universe, and in order to get such a solution from his general theory, he had to modify the basic equations of the theory by adding an extra term, which is known as the *cosmological constant*.

More general solutions of the equations of the theory were obtained by Friedmann (in 1922–1924) and Lemaître (in 1927). These described an expanding universe. Mathematically, an expanding universe is specified by a universal scale factor S which increases with time t. Thus, a distant galaxy is seen at the present epoch t_0, although its image is seen at a past epoch t when light seen today left it. The redshift works out as $1+z = \dfrac{S(t_0)}{S(t)}$. Here $S(t_0)$ is the present scale factor. The function $S(t)$ determines how the universe expands. In the 1920s, the evidence for such a model was not strong, and these approaches were ignored. The situation changed, however, when observations of galaxy redshifts began to accumulate. In 1927, Abbe' Lemaître', a Jesuit priest from Belgium, published his paper in which he used Einstein's general relativity theory to work out models of the universe that showed precisely this behaviour. Two years later, in 1929, Edwin Hubble, working with the 2.5 m telescope at Mount Wilson in southern California, found the aforementioned redshift effect correlated with the distances of the galaxies. The law found by Hubble was

$$V = H. \times D.$$

Here V is the velocity of recession and D the distance of the galaxy. H is called Hubble's constant. Although Hubble gave the first comprehensive analysis of this

PHOTOS: Hubble, Lemaître

effect, we should not ignore the earlier deduction of Lemaître. This was the beginning of a new look at cosmology.

The early universe

The expanding models inevitably lead to the conclusion that the universe started with an enormous explosion (critically named "big bang" by Fred Hoyle) when its density and temperature were infinite. The unphysical nature of this epoch shows that the theory dealing with the physics of the universe is inadequate, and the present attitude of most cosmologists is to ignore this singular beginning and work on the later epochs. Such an approach has led to a typical model of the universe with the following sequence of events.

Big bang at epoch zero

Quantum fluctuations denoting quantum gravity dominated era
Time up to 10^{-41} second

Grand unification breaking into strong, weak and electromagnetic forces, inflation
Time up to 10^{-36} second

Matter–antimatter separation and creation of radiation background
Time up to 10^{-12} second

Formation of light nuclei
Time up to three minutes

Universe became transparent to radiation
Time up to 10^5 years
Formation of large-scale structure
Time up to 10^5 years

This timetable is approximate and subject to change. The first three minutes are believed to be the period when light nuclei (up to atomic weights of 8) were synthesized. For heavier nuclei, the stellar option discussed earlier is followed.

Although this model is currently the most popular one, it has several questions to answer. Some of them are as follows.

1   The background radiation is a very strict black body type. How did it acquire its present temperature of 2.7 K?
2   What determined the matter–antimatter abundance ratio?
3   Do we have a definitive theory of structure formation?
4   What is non-baryonic dark matter exactly?
5   What is the role of dark energy in the overall scheme of particle physics?

Astronomy at other wavelengths

As the physicists following Maxwell realized that visible light is an electromagnetic wave and that outside of the range of wavelengths (approximately 400 to 800 nm), we encounter different forms of radiation ranging from radio waves to gamma rays. So can the astronomer use these other wavelengths for observations? The answer is 'yes,' although there is one problem to solve. That except for radio waves, Earth's atmosphere absorbs the light of most of these wavelengths. So we can build large radio telescopes on the *terra firma*, but for other wavelengths, we need to observe from above the upper layers of the atmosphere. This became possible only after space technology became mature enough to be able to launch telescopes going around Earth *above the atmosphere*. As expected and hoped, these multi-wavelength observations have given very useful information about the cosmos.

Just to give an indication of what has been achieved, we give a partial list:

1   Radio galaxies and quasi-stellar sources
2   Molecules both organic and inorganic in galaxies
3   Gamma-ray bursts
4   Powerful X-ray quasars
5   Ultraviolet sources in the galaxy
6   Microwave background

If one were to single out the most important item from this list so far as astronomy is concerned, the last item is considered to be so.

Dark matter

Astronomers over the years have found that the matter and radiation they observe when they look at stars, galaxies, radio sources, gamma-ray bursts, etc., do not exhaust the list of possible objects in the universe. Since around the 1970s, there are indirect examples of a new kind of matter – 'indirect' because it is not seen through any of the telescopes. But its existence is inferred from the attraction it exerts on the visible matter that astronomers are familiar with. We can thus consider two types of matter (and radiation): ordinary matter that we are made of and observed in astronomy and strange matter that has not yet been detected.

Known as 'dark matter,' it abounds prominently in the universe. If the big bang cosmology is taken as a scale for comparison, then of the total matter in the universe, about 4 percent is the ordinary matter that an astronomer observes, 23 percent is the dark matter and the balance, 73 percent, is ascribed to dark energy for which there is no direct evidence.

The question that arises regarding dark matter, however, is as follows. We saw that the production of light nuclei in the universe happened in the first three minutes after the big bang. Figure 9.5 shows how much the deuterium nucleus is produced in the early universe. The answer critically depends on the density of ordinary matter. It should not exceed a critical value; otherwise, no deuterium will be made. But if all the dark matter were ordinary, its density would not exceed the critical value.

So we arrive at the following conclusion: *there is no deuterium if all dark matter is ordinary. Or dark matter is not ordinary but made of some strange type of particles.*

For the survival of the big bang model, the second alternative is needed. So far, however, no evidence has come for it. Not only in astronomy but also in physics of strange particles, the search for such particles has shown negative results. Yet the implicit faith in this alternative in preference to the first one mentioned earlier is reminiscent of Hans Anderson's story *The Emperor's New Clothes*!

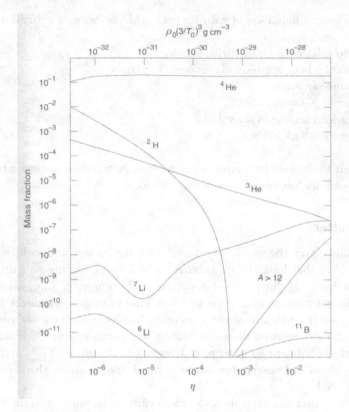

FIGURE 9.5 Deuterium problem. *The deuterium (^2H) problem* can be understood with the help of the graph shown here. The graph shows that the amount of deuterium produced in the early universe drops sharply if the density of ordinary matter (neutrons, protons, etc.) exceeds a certain limit. That is why cosmologists want to believe that most dark matter cannot be ordinary. For, otherwise, the above limit is exceeded.

The microwave background

The microwave background is considered the best evidence for the big bang. George Gamow was a strong proponent of the model and had argued that the early hot era provided a suitable scenario for the creation of chemical elements. This expectation was not borne out except for light nuclei, say those having atomic weights up to 8–10. As it turned out, almost all nuclei with atomic weights 12 onwards can be made inside a star. We have already seen how stars play this role. Ironically, up to the time of writing this account, the early universe and stars play their separate roles in a complementary fashion. Perhaps some bright idea in the future may provide a way for stars to make light nuclei as well.

But apart from nucleosynthesis, another aspect associated with the big bang model is the cosmic microwave background. Two of Gamow's co-workers, Ralph Alpher and Robert Herman, made the prediction that the early radiation-dominated era would leave relic radiation behind, which today would appear in a cool form. There was no theory to lead to a precise value of its temperature but they made an inspired guess of 5 K (that is, −268 C).

This prediction was forgotten, and when Penzias and Wilson found such radiation of approximately 3 K, the Alpher-Herman prediction was belatedly recalled. Subsequent measurements of this radiation showed how remarkably a good fit they were to black body radiation of 2.7 K. Figure 9.6 shows the 1990 measurements of this radiation using Cosmic Background Explorer (COBE) satellite-based instruments.

The problem Alpher and Herman had back in 1948, however, remains unsolved: to determine theoretically the present temperature of the microwave background.

More recently, in the year 2018, a new source became available with the detection of gravitational waves by the Laser Interferometric Gravitational Observatory (LIGO) experiment. The expected signal is typically very small, and so the detector has to be very sensitive to it. Thanks to the technology feedback, the LIGO in its advanced form managed to pick up such signals. Indeed, the next-generation telescopes of various kinds are expected to greatly revise our perception of the universe.

10

A VARIETY OF NEW MATHEMATICS-I

Gauss the prolific mathematician

Sometimes addressed as a prince amongst mathematicians, Karl Friederich Gauss (1777–1855) was an child protégée who almost missed the opportunity of higher education because of the poverty in his family. Fortunately, the reputation of the brilliant boy reached the ears of a local rich person, the Duke of Brunswick, who provided the expenses for his education. From 1795 to 1798, he was at Göttingen University, but while a teenager ager, he solved many problems (previously unknown) that were very difficult. His geometrical construction of a regular polygon of 17 sides using a ruler and compass only is well-known. Indeed, he himself liked it so much that he wished to have it carved on his memorial.

Gauss is well-known for many mathematical results, like his rule for how prime numbers are distributed at large numbers; his proof that every algebraic equation has a root, his discovery of what became known as Gaussian distribution, the method of least squares in statistics, his work in astronomy and instrumentation work in electromagnetism all go to show how versatile his talent was.

A prodigy in a primary school

When the primary school teacher felt that he had need of rest, he decided to give the class an arithmetical problem, which was forbiddingly long. He asked the kids to perform the following sum:

$$1 + 2 + 3 + \ldots + 97 + 98 + 99 + 100 = ?$$

In short, the (poor) kids were asked to perform a hundred summations. The teacher was new to the class, and he felt that by asking them to make such a long series of

DOI: 10.4324/9781003203100-11

sums he would create a tremendous impression in their minds about himself. He was sure that while doing a hundred sums, the kid would stumble somewhere and give a wrong answer.

The students used slates for writing and went on their 'summation' to find the answer. There were all kinds of stumbling blocks, but the teacher was gloating in his mind that he could find such a time-consuming problem – until his roving sight fell on the boy in the first row.

He was not working.

The teacher felt here was an idle fellow who needs to be energized. He called the boy to the teacher's chair with his slate. There was a single number written on it: 5,050.

"What is this for?"

To the teacher's query, the boy replied, "This is the answer you are asking for, sir."

The teacher knew that that was the case. But how did the boy get that answer? He asked the boy to explain.

"Well, sir! We have to sum numbers from 1 to 100. We make pairs from these: [1, 100], [2, 99], [3, 98] and so on . . . pairing the first and last numbers. Each pair adds up to 101. There are altogether 50 such pairs. So it follows that the sum of all such pairs is 5050."

History tells us that the boy grew up to be the most famous mathematician Karl Friedrich Gauss, and his teacher in the primary school was Büttener.

Gauss on astronomy

Gauss is better remembered for his work on pure mathematics than on astronomy. In fact, when he was looking for an academic position that could supply his livelihood, he felt that with mathematics being a 'useless' subject, he would not be able to support himself. So he was happy when offered the directorship at Göttingen Observatory and professorship of astronomy. He held that position till the end of his life.

One of his brilliant works was in relocating the dwarf comet Ceres, which had been seen earlier and had later been 'lost' to observation. The mathematical techniques employed by Gauss were very much responsible for locating it. This work later led him to work on data analysis and statistics. His work on normal distribution has led to that distribution being called "Gaussian."

New lamps (geometries) for old

We shall see how science changed drastically with the addition of relativity and attempts to understand nature seemed to fail in the process when applied to understanding natural phenomena. The new requirements led mathematicians to produce new types of mathematics. Although the mathematicians would deny that their subject was producing new branches to meet the specific needs of scientists, it just happened that way, as we shall see now.

We will start with geometry. As we saw, Euclid's geometry had served science well, and there was every reason to believe that it will continue to do so. In our earlier discussion (p. 7), we mentioned that Euclid's parallel postulate (Figure 1.2) had led to unsuccessful attempts to *prove it* so that it may no longer remain an axiom. It took a lot of effort to resolve the issue, as late as the nineteenth century.

Δ ABC is drawn with base BC extended to point D (See Figure 10.1). Using the parallel postulate, CE is drawn parallel to AB. Parallel line properties show that ∠ ABC is equal to ∠ ECD and ∠ BAC equals ∠ ACE. So we have ∠ ABC + ∠ BAC + ∠ ACB equalling ∠ ECD + ∠ ACE + ∠ ACB = 180°. Thus the parallel postulate is needed to claim that any triangle has the sum of its angles equalling 180°.

Notice that the crucial step in the proof requires the parallel postulate. So, if the postulate were not there, the proof could not work.

Hence, if we abandon the parallel postulate, we will have a geometry in which the three angles of a triangle *do not add to 180°*.

At this stage, an incredulous reader may ask, "Where on Earth is this possible?" The answer is, "On Earth itself."

To see how this is possible, imagine Earth as a perfect sphere. We want to draw a straight line connecting any two points P,Q on Earth. Defining the straight line as the path of least distance (the same definition as on a plane!), we use a stretched elastic band to connect those two points. The band will take up the shortest route from *P* to *Q*. Now we imagine a hiker starting on a long, long hiking tour beginning from the North Pole. He or she takes the Greenwich Meridian and walks along it to the equator. Upon reaching it, he or she turns left. The hiker then walks a quarter of the circumference of Earth to arrive at 90° longitude. He or she then walks on this longitude to arrive back at the North Pole. There, the hiker turns left so as to get back to his or her starting position. The hiker will describe walking in a triangle with three angles each equal to 90° (recall that the hiker made three turns to the left; Figure 10.2).

Clearly, something has changed! What? The answer is that the parallel postulate does not operate on the surface of the sphere. There are no parallel lines here: any two straight lines drawn on Earth must intersect when extended fully. We will find

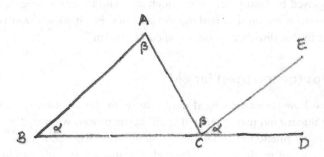

FIGURE 10.1 Euclidean theorem. A theorem in Euclid's geometry is shown using the parallel postulate.

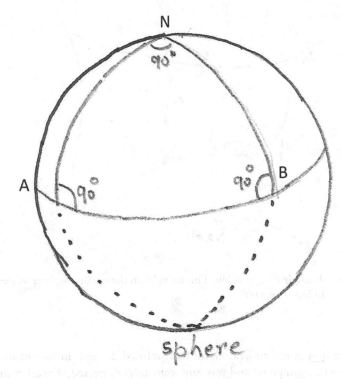

FIGURE 10.2   Triangle on Earth Point N at the North Pole; A, a point on the equator where the Greenwich Meridian intersects it; and B, a point on the equator at 90° meridian. All three lines of this triangle are the shortest distance between the three vertices N, A, B. But the three angles of this triangle add up to 270°.

that all triangles drawn on the spherical Earth will have their three vertex angles adding up to more than 180°.

So we have seen that another geometry can exist not subscribing to the parallel postulate. As a second hypothesis, can we find a surface such that any triangle drawn on it will have its three angles adding up to *less* than two right angles? The bridle on horseback is one such surface. We can draw on it an infinite number of parallel lines as shown in Figure 10.3.

Non-Euclidean geometries

The mathematician Karl Friederich Gauss, described earlier, had claimed to have discovered non-Euclidean geometries. Given his versatility and past creativity, such a claim would have stood the test of creativity. However, unfortunately, this was not the case. The credit for creativity goes to Janos Bolyai (1802–1860) and to Nikolai Lobatchewsky (1792–1856). A Hungarian by birth, Bolyai's father was himself a bright mathematician and one-time student of Gauss. But when Janos applied to

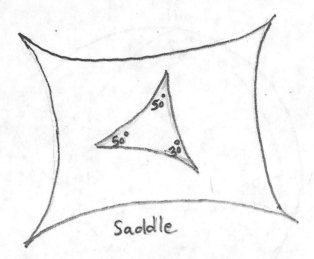

FIGURE 10.3 Triangle on the saddle This triangle on the saddle has three angles adding to less than 180°.

him to take him as his student, Gauss flatly refused. Indeed, in his old age, Gauss had become bad-tempered and was (unfortunately) suspected of plagiarism.

Janos then joined the army for his livelihood. He learnt as many as nine languages, including Chinese and Tibetan, that he picked up when he travelled on assignment to those countries. All the time, however, he was obsessed with the non-Euclidean geometries, despite being warned by his father to keep away from the topic at the risk of going mad! After trying to disprove the parallel postulate, he finally realized that it provided the route to non-Euclidean geometries. He, therefore, spent his time studying those, and he wrote a lot on them. But he did not bother to publish his research, with the result that the only publication to his credit was the 24-page introduction he wrote for his father's textbook. He got isolated and even suspected of going mad. His work of some 20,000 pages was published posthumously.

The other author of original work on non-Euclidean geometries was Lobatchewsky born in Kazan, Russia. Like Bolyai, he too went after disproving the parallel postulate and in the process ended up discovering non-Euclidean geometries. His work first came out in 1826, to be published in 1830. Both he and Bolyai are credited with opening up the exciting work on non-Euclidean geometries.

Differential geometry

The subject of calculus invented by Newton and Leibnitz had turned out to be very useful in various branches of mathematics and physics. Gauss was the first

PHOTO: Gauss

to appreciate that it could be used in geometry. For example, a lens with curved surfaces is used in various ways. Light passing through it is bent depending on the curvature of the lens. Thus a scientist interested in a specific type of lens will need to work out how the curvature on the two sides of the lens is built up.

In general, problems such as this will need calculations. These are carried out with the combination of calculus and geometry of surfaces. A systematic approach to this usage is through the branch of geometry called *differential geometry*. Gauss was the chief contributor to this topic.

Although Gauss did not give much help to the up-and-coming young mathematicians, one of the few who did benefit from his help was Bernhardt Riemann (1826–1866), who came from a poor family in northern Germany. Because of his brilliance, he gained admission to the well-known Göttingen University. Gauss in particular recognized his brilliance and encouraged him. Since the public image of Gauss was not particularly happy (because of incidents relating to non-Euclidean geometry), this interaction was something of an exception! Riemann worked on a specific type of differential geometry, which he rightly expected to be very useful. This geometry is known after him. Later, when working on his theory of general relativity, Einstein found *Riemannian geometry* to be very useful.

Apart from his work on geometry, Riemann also made an important contribution to the theory of numbers. He discovered a function which was closely related to the distribution of prime numbers. The function is known as Riemannian Zeta Function. It will take us too much into technicalities to describe this relationship. However, we will briefly return to it at a later stage.

The seven bridges of Königsberg

The small problem raised by simple citizens of a small Prussian town indeed turned out to be much more complicated than seemed earlier. A river called Pregel flowed through the central part of the town but had two small islands in it. As shown in Figure 10.4, there were seven bridges over the river, all within walking distance of

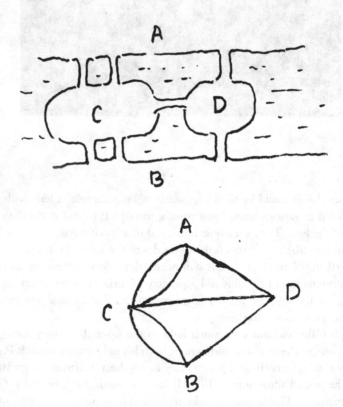

FIGURE 10.4 The seven bridges problem. The way the seven bridges were distributed is shown. For the sake of argument, the equivalent graph is drawn next.

one another. The townspeople liked taking a walk on some or all of those pictur-esque bridges. The observant amongst those noticed that although it was possible to take all the bridges in one go, it did not seem possible to cross all the bridges so that no bridge was crossed more than once.

The year was 1735. The town of Königsberg had a local distribution of bridges connecting the opposite banks of river Pregel and also two landmasses in the river. Figure 10.4 gives an overall idea of the local geography. The locals often went for walks and tried various combinations in an attempt to cover all seven bridges once and only once. They did not succeed. They then approached Euler, who was one of the foremost mathematicians in Europe. "I promise," he said, "either to show how it can be done or give a proof that this cannot be done."

So Euler thought and thought and finally came up with a proof that what was being asked for was not possible. Here we try to give a simplified version of Euler's proof.

The following version of the seven bridges makes it easier to come to the proof. Suppose we shrink each bridge to a point. Then the original diagram changes to what is shown next to the original figure. Now we have a 'graph' with four vertices connected by seven edges.

Euler argued that in Figure 10.4 we have to traverse on all of the four vertices. But no edge is traversed twice. In such a trial, we use one edge to go to one vertex and another to leave the same. In general, this has to happen at all vertices. To visit a vertex and leave it, the vertex must have an even number of edges. The only excep-tion is at the beginning and end of the journey, which can have odd edges. This requirement is not satisfied by the four vertices. All of them have an odd number of edges (3,3,3,5). So Euler's answer was that the problem was not solvable.

But we can reveal that his reasoning enabled Euler to begin a new branch of mathematics, which is today called *graph theory*.

The four-colour map problem

It was the year 1856, when Augustus De Morgan (1806–1871), a mathematician at the newly established King's College, London, wrote to his friend the Irish mathe-matician William Rowan Hamilton (1805–1865), mentioning what was to become well-known as the "Four Colour Problem" (Figure 10.5). It had been suggested to De Morgan by his student Frederick Guthrie who had got it from his elder brother Francis who later became the professor of mathematics at Cape Town University.

The problem was simple to set but very difficult to solve: which is why it travelled amongst the mathematical community causing ripples, with many false proofs! The problem was to do with colouring maps. When we look at a geo-graphical map, we see countries painted in different colours such that neighbouring countries (i.e., those sharing a common boundary) have different colours. If so, how many colours are needed in the least? It is not too difficult to show that five colours are sufficient for this purpose. But a little more work will suggest that four colours may be minimally needed. However, the mathematical proof was not so

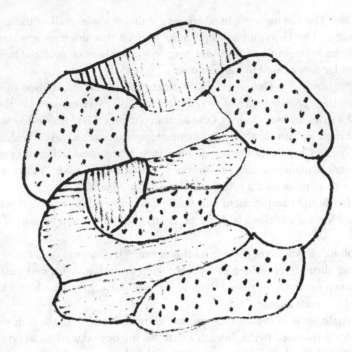

FIGURE 10.5 Four colour problem. A typical map is drawn and neighbouring countries are coloured differently. Four colours can do the job here. Is that the
case for all maps? Are four colours always enough?

easy to obtain. In fact, to prove that four colours are enough was one of the difficult
problems in mathematics.

The delusive proof tempted many mathematicians and other intelligent persons
to try for a proof, but all fell by the wayside in some cases with subtle errors that
had escaped the attention of many who checked the arguments. In the last century,
however, two mathematicians, Wolfgang Haken and Kenneth Appel at the University of Illinois, found a way of classifying maps in which the errors or cases missed
were checked by computers. As the number of cases to be checked ran up to several
thousand, there did not seem to be any manual way of handling the problem.

11

A VARIETY OF NEW MATHEMATICS-II

Short-lived algebraists

While we found that many new possibilities existed for geometries once we allowed a breakdown of the parallel postulate, likewise, for algebra, we open new algebraic systems once we permit the relaxation of one postulate. For example, we could relax the condition that all products are commutative. In short, if we define the product of two members X and Y of a set as XY, we could also write it as YX.

Now, it may be the case that X and Y are not commutative; i.e., we do *not have* XY = YX.

We can define the elements of an algebraic system as operators. For example, consider operators referring to putting shoes on and putting socks on. Let X = putting socks on and Y = putting shoes on. Then YX denotes the operations of putting on socks followed by putting on shoes. This makes sense! But what about XY? It is clearly different from YX. (In this example, any operator acts from the right.)

In that case, you have a different algebraic system: different from what we used in our school algebra.

At a more technical level, mathematicians define specific systems called groups, rings, etc. In all such cases, we need to specify what is the condition of commutativity. Do we really need algebras that use non-commutative multiplication? We will later describe one case where non-commutative algebra is actually needed. Here we will describe two algebraists who were brilliant but were short-lived.

Evariste Galois (1811–1832)

Born in France, Galois was interested in the works of contemporary mathematicians like Lagrange and Legandre. He handled questions that were quite difficult, but he was otherwise not known as a bright boy at school. He failed the entrance

DOI: 10.4324/9781003203100-12

PHOTO: Galois

test for the celebrated school Ecole Polytechnique in Paris. But he carried on with mathematical works regardless of what the prescribed topics were.

Galois used the notion of group theory to solve certain fundamental problems. To understand their nature, recall the so-called *quadratic equation* such as

$$Ax^2 + Bx + C = 0.$$

Here A, B, C are given constants, and we are required to find the solution – that is, the values of x which make the left–hand side equal to zero. The problem is solved at the pre–undergraduate level and the solutions are two as follows:

$$X = \frac{1}{2A}\left\{-B \pm \sqrt{B^2 - 4AC}\right\}$$

Note that the solution is expressed in terms of the given constants. The question arises if a similar result can be claimed for an equation of third degree. The answer is "yes." What about the fourth degree? The answer is still "yes." But there the success ends! For fifth and higher-degree equations, this is not possible. Why?

Galois got a solution to this long-standing problem with his approach using group theory in algebra. It is now well understood why a fifth or higher-degree algebraic equation cannot be explicitly solved in terms of its given constants.

Galois's method also helped explain the "why" of the following questions:

Why is it not possible to trisect any given angle with a ruler and compass (that is, divide a given angle into three equal parts)?

Why is it not possible to construct a cube of twice the volume of any given cube with a ruler and compass – that is, given a cube of side 'a' (with volume a^3) to construct a cube of volume $2a^3$?

Why is it not possible to construct (with a ruler and compass) a square of area equal to the area of a given circle?

Thus we have in Galois an up-and-coming young mathematician who would go far in his mathematical research. However, fate decreed otherwise. There was another aspect to Galois's interests.

After the French Revolution, Galois became a very dedicated Republican and spent a lot of time politicking especially in view of the conditions prevailing in France in the reign of King Louis Phillippe. In these circumstances, Galois had got into an argument with someone, which resulted in his being challenged to a duel. Being aware that it was not possible to get out of a challenge whose result could be his death, Galois sat up late the previous night and wrote out all his mathematical findings. As he expected, he was killed in the duel but his work survived and posterity saw how great his work was. It was the loss of his life to a trivial cause that posterity would regret.

Niels Henrick Abel (1803–1829)

Abel, a contemporary of Galois, had a different temperament altogether. Unlike the impulsive responses of Galois, Abel had a well-balanced role cut out for him largely because he had the responsibility of family: parents and seven children. When his father died at the age of 48, Abel took on the responsibility of maintaining the family without batting an eyelid. Also, it took him longer to realize that he liked mathematics and more so that he had the seeds of a mathematical genius. He had the good luck of having an understanding teacher, Bernt Michael Holmboë, who though with no pretensions to mathematical genius did understand that his pupil was an exceptional boy who would add considerably to the work of previous mathematicians like Newton, Euler, Lagrange, etc. Abel, though a teenager, was well aware of some of the errors in the proofs of these stalwarts in areas like the binomial theorem, infinite series, etc.

His own work in many ways ran parallel to that of Galois. While describing his own work, Abel set up two major aims:

To find all equations of any given degree which can be solved algebraically.

To determine if any given equation is solvable or not algebraically.

PHOTO: Abel

Teacher Holmboë felt and Abel agreed with him that Abel should go on a European tour to meet and discuss his work with the leading mathematicians. They applied for funds, but the country (Norway) was passing through a postwar period of poverty. But he did get some support though since his reputation was a help.

In Berlin, he met August Leopold Crelle who was to be of great help to Abel. Crell brought out a journal of high quality to which he got Abel as a contributor.

Incidentally, at different times, both Galois and Abel had sent some of their work to the French mathematician Augustin-Louis Cauchy for his views, and he managed to lose it!

12

MAGICAL DECADE AT GÖTTINGEN

The Camelot years at Göttingen

The period 1920–1932 was the golden age at Göttingen, a small town in Germany. This was the time when distinguished seniors, as well as young hopefuls, gathered in this town and tried out ideas and speculations on one another. Since it was known that brilliant ideas would be presented in lectures by scientists and mathematicians, visitors flocked to this city, both to hear them and to try out their own speculations. For it was a no-holds-barred atmosphere, and in the summer season, the not too wide roads were jammed by walking or cycling thinkers. (Motor cars were few and far between.)

Indeed, these summer visitors to Göttingen were responsible for sustaining the economy of that tiny city. Knowing this fact, the shopkeepers permitted advances to the impecunious young men who may not have ready money to pay but were expected to be able to settle their outstanding bills in the end. The story went around that one young student offered a performing bear in lieu of cash, which the shopkeeper rejected saying, "If I accept that, I will be tied up like this animal." Amongst the distinguished were Gauss (reputed to have visited in the past), Felix Klein, Runge, David Hilbert, the mathematicians, and Paul Dirac, Landau, Robert Oppenheimer, amongst the scientists.

It was Felix Klein who had initiated the invitations to the overseas talent, even though he had controversies with some of them. Some were not happy that many of these invitees were applied mathematicians, while others were known for their rudeness. One such personality was Hilbert, whose lectures always filled up the auditorium. It was guaranteed that he would say something provocative or suggest new ideas in his lecture.

A problem that Hilbert never attempted or even tried to attack was Fermat's last theorem. Hilbert had an amusing reason for doing so. The problem had been

DOI: 10.4324/9781003203100-13

PHOTOS: Hilbert, Dirac, Oppenheimer

advertised with a big reward, and it was stated that until the prize money was given away to the award winner, it would be deposited in a bank with the proviso that the interest of that capital would be used for covering expenses of select visitors. So Hilbert used to say, "If I attempt solving the problem and succeed in finding its solution, this source of visitor programme would be lost." So he would not like to kill the goose that laid a golden egg!

When Einstein was nearing the end of his formulation of general relativity theory, Hilbert had stepped in and found an elegant and formally more convincing approach to derive the basic equations of the theory. There are many such contributions by Hilbert.

Every Thursday promptly at three in the afternoon, four mathematicians from the Mathematics Institute, Klein, Runge, Minkowski and Hilbert, used to meet on the verandah of Hilbert's house. Sitting there, one could see the garden. There was a blackboard and chalk ready for use. Sometimes they would step out and walk to the Keher Hotel to have coffee there. To an outsider, the occasional bursts of laughter would convey the information that not all being debated was serious!

In the auditorium building, Klein had provided a reading room with a small library containing research journals. It was all quiet here, but in the neighbouring room, there would be a lot of noise from arguing scientists and mathematicians. One provocative comment of Hilbert was, "Physics is too important to be left to physicists. They can't cope with it." In 1921, Hilbert, Franck and Max Born had worked together here on quantum mechanics. Hilbert would sometimes start his lecture with this question to the audience: "What is exactly an atom: would somebody here tell me?" This would trigger off a class-wide discussion.

An important component of the visitors came from the United States. The United States was quite behind Europe in research and teaching of maths and science. To catch up, there were several visitorships enabling the bright young Americans to come to Europe for higher studies. Many of them would end up in Göttingen and amongst them was a highly talkative young man named Robert Oppenheimer. Although very bright, his colleagues once complained to the authorities to prevail on him to talk less!

Oppenheimer (shortened to Oppie) was very well-read and even knew the Sanskrit language. Later, he became the leader of the Manhattan Project for making the atom bomb. It is said that when the first experimental bomb was exploded, the enormous brightness produced reminded him of the verse in the Hindu scripture *The Bhagavad* Gita, making a comparison with a thousand suns appearing simultaneously in the sky.

The American visitor programme brought valuable dollars to Germany, and so these visitors were always welcome. As luck would have it, both Oppie and Dirac were housed in the same building. Dirac was known for talking less (it was said that Dirac talked one sentence in a whole year!). So much so that Dirac asked Oppie, "You are a poet, I hear. How do you manage it while doing physics? In physics things are simplified whereas in poetry things are made more difficult to understand."

The fame and success achieved by the Göttingen school did not continue beyond 1932. The rise of Hitler and the Nazi culture put a stop to most academic activities. Many scientists and mathematicians migrated to the United States, and the visitor numbers at Göttingen sharply declined. The Camelot era of Göttingen came to an end.

13

FIRST SHOCK

Relativity

A feeling of complacency had started to develop amongst physicists towards the end of the nineteenth century – a sense that they were approaching the summit of knowledge and that a stage would come before long when most of nature's outstanding mysteries would be resolved, leaving a few details for mopping up exercises. In fact, a few mysteries that were around turned out to be misleadingly simple. We will describe two of them, both relating to the speed of light.

The Michelson-Morley experiment

Consider the following argument. A man wishes to row in a river which is flowing with speed v in the easterly direction shown in the figure as labelled OX (Figure 13.1). Suppose that in still water his rowing speed would be c so that in the flowing river his effective speed would be $c + v$ in the direction of flow and $c - v$ in the opposite direction. Suppose he rows a distance l in the direction of flow and back the same distance. What will be the duration of the entire exercise? It will be

$$T_{EW} = l/ (c + v) + l/(c - v) = 2lc/(c^2 - v^2).$$

Here the suffix EW identifies the east-west direction of the rower.

Now we look at a second rowing exercise: the rower rows in such a way that he is always moving perpendicular to the river flow. To achieve this, he has to choose an oblique direction in which he rows such that he effectively moves in the north-south direction. A simple vector algebra tells us that his speed in the north-south direction is $(c^2-v^2)^{1/2}$ so that he makes a north-south trip of length l and back. From these calculations, we deduce that the east-west trip takes a longer time than the

DOI: 10.4324/9781003203100-14

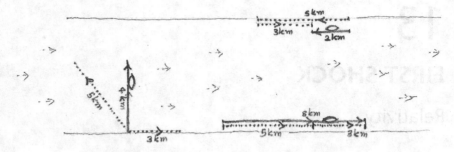

FIGURE 13.1 Rowing problem A man rowing along a flowing river finds that his effective rowing speed is increased: it is the sum of his speed in still water and the speed of river flow. For the return journey, his effective speed is reduced by the river speed. The corresponding journey perpendicular to the river flow is as stated in the text. The final time taken for the new journey is slightly less than that in still water. A similar result was expected when journeys of light vis-à-vis aether were observed. As mentioned in the text, the aether was hypothesized as a medium for the propagation of light. Despite very accurate instruments Michelson and Morley failed to detect any.

north-south one. If we measure the extra time taken for the east-west trip, we can determine the river speed.

$$T_{NS} = 2l/(c^2 - v^2)^{1/2}$$

Now keep this experiment in mind while arranging a more sophisticated experiment. This experiment uses Michelson's interferometer, which compares the speed of light in two perpendicular directions. Since Earth moves in its orbit around the Sun at the rate of 30 km per second in the east-west direction, the original experiment was planned to measure this speed against a stationary medium called aether. The realization that the light moves in the form of an electromagnetic wave, as shown by Maxwell's equations, it was nevertheless assumed that the waves required a medium to travel. We refer the reader to books on relativity for the details of this experiment. Here we note that like the rower measuring the speed of the river, Michelson and Morley were expecting to measure the speed of aether.

But they did not succeed in establishing the expected excess time light would take in the east-west direction. Although the expected answer was quite small, the sensitivity of the interferometer was much more, and their conclusion was that the speed of aether was zero.

From Maxwell to Einstein

While the Michelson and Morley experiment tried to determine the speed of aether relative to that of Earth, an entirely new perspective was brought to bear on the issue by a clerk in the Swiss patents department. Although in a civil job, the young man named Albert Einstein was deeply interested in theoretical physics. In particular, he was concerned with Maxwell's electromagnetic theory.

By the 1860s, as a result of a large number of experiments by Volta, Ampere, Faraday and many others, the electromagnetic theory had been built up as a complete theory. The build-up had been stage by stage and the final 'brick' in the structure had been put in by James Clerk Maxwell. We briefly referred to it in Chapter 8. The major conclusion of the theory was that the electrical and magnetic disturbances travel in an interrelated fashion with a speed no different from

PHOTO: Einstein

the speed of light. As we saw in Chapter 8, Hertz demonstrated this conclusion experimentally.

However, when Einstein studied the electromagnetic theory, he ran into a practical problem.

The equations lead to the conclusion that the electric and magnetic fields travel as a transverse wave with the speed of light which we will henceforth denote by c. Thus an observer at rest with respect to the source of light will measure its speed to be c. So if another observer travelling with speed v towards the same source measures the speed of light, he should get the answer as $c + v$. This answer is based on the basic laws of motion laid down by Newton and is consistent with our everyday experiment.

But not so in the previous case. Maxwell's equations as written down by the moving observer would be the same as for the stationary observer. So the former should get the answer as c and not as $c + v$. Since velocities are measured by dividing the distance covered by the duration of motion, there is evidently some discrepancy. Einstein then worked out the revised rules for space-time measurements. The same revised rules turned out to explain the 'null' result found by Michelson and Morley.

The previous result was deduced by Einstein from the requirement that the Maxwell equations are the same in form for any two observers in uniform relative motion. From this result, Einstein could conclude a more universal result – namely, that these observers would find the same form for any basic scientific law. Expressed in this form, he called the underlying theory the *special theory of relativity*. The word 'relativity' describes the invariance of physics for any two observers in uniform *relative* motion: the significance of the adjective 'special' will become clear soon.

The mathematical rule connecting the space and time coordinates of two observers in uniform relative motion is called *Lorentz transformation* so named after a contemporary physicist Henry Lorentz who had proposed it in connection with his ideas on contraction of a moving body. These same mathematical relations turned up in the ideas proposed by Einstein. The concept of observers in uniform relative motion is assumed to describe the idealized observers on whom *no force acts*. By Newton's first law of motion, which is often called the *law of inertia*, such observers are called *inertial observers*.

The special theory of relativity

As we see, the special theory of relativity describes a symmetry between any two observers in uniform relative motion. As such, in the early days, this idea led to apparent contradictions, puzzling even professional physicists, such as the following example. If an observer A carrying a rod of length L passes with speed v another observer B, B will note the times when the two ends of the rod cross him. Using the formulae of relativity, he works out the interval between the two ends crossing B and that multiplied by A's speed he gives the length of the rod as

$$L \times (1 - v^2/c^2)^{\frac{1}{2}}.$$

Thus B concludes that A's rod is not of length L but is shrunk by factor $\gamma = (1 - v^2/c^2)^{1/2}$. But then a similar experiment by A will tell him that B's rod is similarly shrunk. So what is the truth? A more puzzling version of the example goes under the name "twin paradox." This envisages two twins A and B on Earth of age, say, 20 years. On his 20th birthday, twin A leaves on a cosmic tour travelling in a fixed direction with speed $v = 3c/5$. He travels as per the time kept on Earth for time T and then decelerates and comes to rest and reverses his motion for time T, returning to Earth when his twin brother B has aged by $2T$. *How much has A aged in this exercise?* The formulae tell us that A's watch runs at the rate

$$f = (1 - v^2/c^2)^{1/2} = 4/5$$

times the standard clock on Earth followed by B. *Thus when he reaches Earth, A will be younger than B by the time interval* $4/5 \times 2T$. Thus if $T = 10$ years, when A is back on Earth, his age will be 36 years while B will be 40 years old.

The paradox arises if we argue that as seen by A, B goes on a similar journey and so would be likewise younger than A. So what is the real solution? Arguments like these made it hard for people to accept the special theory of relativity. The difference between A and B is that B is an inertial observer whereas A accelerates and decelerates. For such observers, the Lorenz transformation does not apply. The formulae for observers like A are different, and they confirm that A will indeed be younger than B when he returns. But as more and more experiments demonstrated the applicability of the theory, it gradually gained acceptance. The concept since Newton's assumption that time is 'absolute' and the same for all observers was shown to be wrong. The phrase 'time dilatation' came into physics as a result of special relativity (SR).

A striking demonstration of time dilatation was given by the observation of muon particles in cosmic rays. In the laboratory, a muon is not stable and lasts for a decay time of approximately 2 microseconds. The decay is into an electron, a muon neutrino and an electron antineutrino:

$$\mu^- \rightarrow e^- + \bar{\nu}_e + \nu_\mu$$

During their lifetimes, they could not have travelled more than 600 m (2 microseconds × c). *But despite this short lifetime, muons are seen to be present in the cosmic rays for at least ten times longer.* This is because of time dilatation. At a speed of 0.995 c relative to us as observers on Earth, the muons last that long.

Aspects of SR

When Einstein started examining the consequences of SR, he found them very unfamiliar indeed. Let us look at a few effects that are hard to understand.

Consider the result of SR that light velocity is the same for all uniformly moving observers. To see the effect, imagine a train moving at the speed of 300 km per hour. On a parallel highway, two observers are moving, *A in the direction of the train* and *B opposite to it*. Let us assume that both cars are moving at the speed of 100 km per hour. Driver *A* will find that the train overtakes him relatively slowly (with a speed of 200 km per hour). Driver *B* will find that the train flashes past the driver at the speed of 400 km per hour.

While this appears reasonable, what could one say if the speeds of the train seen by the two drivers seem the same? If we replace the train with light, the speed of light measured by the two observers will be the same.

According to SR, the speed of light should appear to be the same to all observers, *whatever their speed. Indeed* the previous requirement becomes hard to understand. Other similar effects are the following:

1   A rod is moving in a lateral fashion, being carried by an observer *A*. Another observer *B* observes the rod from his rest frame. Suppose this observer sees the rod at length 1 m. When they compare notes, they find that the rod length measured by *A* is longer. A similar result is seen if the rods are interchanged – the rod seen at rest has a longer length.
2   A similar result follows if one compares moving clocks. The clock at rest would register a longer time.

If we have two inertial observers in relative motion, how do they relate their space-time measurements? In Newtonian dynamics, the rules for this were those used from the time of Galileo. Different rules were needed to accommodate the aforementioned type of effects. As it happened, the rules were proposed by Lorentz on empirical grounds. They are the *Lorentz transformations*. It was seen that to understand the 'strange' effects, these new rules were needed. Using these, Einstein found that the mass of a moving body is not fixed but increases with velocity. The extra part of mass comes from the energy of motion. Thus Einstein was able to arrive at the famous relation $E = Mc^2$.

Conflict with Newtonian dynamics

When Einstein looked at the Newtonian laws of motion and of gravitation in the light of SR, he saw that there were serious conflicts. We will not go into the details but only highlight some basic issues. First, imagine that by magic (!) the Sun is removed from the planetary system. What will happen? Because Newtonian gravitation is instantaneous, Earth will immediately notice the effect: it will no longer move under the inverse square law force but by Newton's first law of motion, take off in the tangential direction to continue moving with the same speed in the same direction. However, because light travels from the Sun to Earth in approximately 500 seconds, we would see the Sun disappear after that time. A more consistent situation would have been if the gravitational effect had also taken 500 seconds to reach Earth. In short, paradoxical situations could be prevented if gravitational force also travelled subject to the speed limit of SR (Figure 13.2).

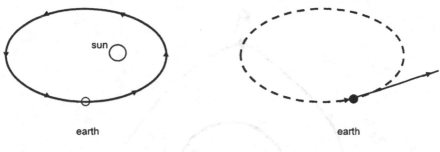

If the Sun vanishes

FIGURE 13.2 Earth–sun motion if gravity vanishes. The Earth goes around the Sun in an elliptic orbit. If the Sun were to vanish by magic, the Earth would move (according to Newton's first law of motion) in a straight line, which would be tangential to the elliptic orbit, as shown here.

Secondly, the inertial observers which play the key role in SR do not exist because there is no location in the universe free from gravity. Thus, the application of SR is itself of questionable validity.

Problems like these made Einstein wonder if both Newtonian gravitation and SR need to be modified radically to achieve consistency. Indeed, he eventually arrived at a theory which combined both dynamics and gravitation. Known as *the general theory of relativity*, this formulation required giving up the tacit assumption that the space-time measurements require Euclidean geometries.

In Chapter 10, we encountered this notion – at least as a new kind of geometry. As products of the weird imaginations of mathematicians like Bolyai, Lobatchewski and Gauss, these geometries were not expected to have any application to real life. Einstein proved the belief to be wrong. We will not go into the details of the exercise but try to understand this strange concept.

The general theory of relativity

The problem of inconsistency between SR and Newtonian physics was sought to be resolved by Einstein using a radical approach. He started with the assumption that in the absence of gravitation, space-time follows Euclid's geometry, whereas in the presence of matter and energy, there will be gravitation and hence *non-Euclidean geometry*. The notion of non-Euclidean geometry was briefly discussed when we recalled that mathematicians had already thought of such geometries not because of their physical relevance but because of aesthetic and completeness criteria. Einstein adopted Riemannian geometry as the underlying geometry of space-time in the general case. Einstein's equations set down rules of how to relate the parameters of the space-time geometry to the presence of matter and energy. He called this formulation *the general theory of relativity*.

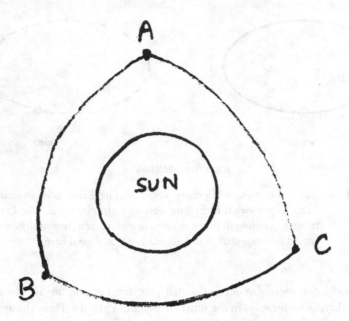

FIGURE 13.3 Triangle around the sun. The thought experiment described in the text would show the triangle has three angles, adding up to slightly more than 180°.

   The way Einstein brought science and mathematics together can be understood by this thought experiment. Imagine three observers *A, B, C station themselves around the Sun* in such a way that light rays joining *A* to *B*, *B* to *C* and *C* to *A* and touching the Sun's surface are equal in length. Indeed, we would have called the triangle *ABC*, an equilateral triangle, if it had conformed to the Euclidean notion of such a triangle (see Figure 13.3). Instead, a very sensitive angle measurer would tell us that the sum of the three angles of the space triangle exceeds 180° by a very tiny amount

$$21 \times 3^{1/2} / 4 \times \{GM/Rc^2\},$$

where *M and R* are the mass and radius of the Sun.

   This type of result shows that gravitation bends light rays thereby answering the query Newton had raised (vide Chapter 6).

Why do we need general relativity?

Let us now ask, as many must have done in the decade, Why do we need general relativity? The theory henceforth referred to as GR seemed to require many complex ideas. Could one not do without them?

Let us start where SR left off. In a sense, Maxwell's equations deducing the invariance of electromagnetic equations forced the invariance with respect to the Lorentz transformations. Given this result, it became necessary to modify dynamics, including Newton's laws of motion. And that gave rise to the famous equation $E = Mc^2$.

But the path chosen by him was not a slight perturbation on how Newton had looked at gravitation. It was a wholesale change! He used as a clue the observation that gravitation is not removable! That is, he associated gravitation with *a property of permanence*. His masterstroke was to link gravitation with a new kind of geometry. Quantitatively, he set up a series of equations relating to the parameters of space-time geometry.

The year 2015 marked the centenary 2of Einstein's *general* theory of relativity, regarded by many as an intellectual effort of the greatest originality. Einstein had a decade earlier shocked the scientific establishment with his highly radical *special* theory of relativity. Although that too had been a great feat of thinking, Einstein had been aware that there were still problems associated with the coexistence of special theory and the Newtonian law of gravitation. Searching for a more comprehensive theory, he realized that he needed to empower himself in mathematics. Despite help and guidance from his mathematical friend Marcel Grassmann, it took Einstein ten years and some wrong turns on the way to arrive at a satisfactory framework for the general theory of relativity.

The mystery of the perihelion of Mercury

However, there was a historical issue that provided support to GR. Recall how Neptune was discovered. Its presence had caused a local discrepancy in the motion of planet Uranus. In the mid-1840s, both Adams and Leverrier had noted the discrepancy and used it to deduce the existence of the planet Neptune. However, the same trick did not work when the planet Mercury began to show a problem. The point in the orbit of Mercury called the *perihelion*[1] was steadily moving, as shown in (Figure 13.4). The shift was as small as 42 arc seconds per century.

In the 1860s, Leverrier was known as a distinguished astronomer, and he tried to solve Mercury's problem by assuming a new planet called Vulcan, closer to the Sun. He came to this conclusion but could not support it by finding the new planet. When GR came into the field, calculation showed a precession in Mercury's perihelion exactly as calculated by GR.

We can see a moral of how science may progress. There can be a discrepancy between theory and observation. The agreement between the two can come from an error in observation. This happened when Uranus strayed from its expected path. Why did it do so? Because a 'hidden' planet was causing the disturbance. This reason however did not work for Mercury. To explain its anomaly, one needed a theory different from Newton's. Science has different ways of revealing the truth!

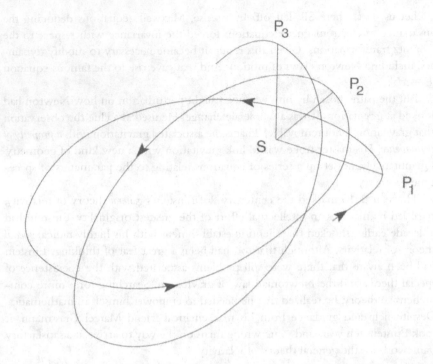

FIGURE 13.4 Perihelion of Mercury orbit. The perihelion of Mercury (the point closest to the Sun in its orbit) slowly precesses $P_1 \rightarrow P_2 \rightarrow P_3$. This effect could not be explained by Newtonian physics but followed naturally from GR.

Where the bureaucrats feel at home

One of our puranik tales describes the experience of King Kukudmi who had a beautiful daughter named Revati. As she approached marriageable age, several young men started approaching her father with marriage proposals. With the natural desire that his choice should be right, he finally decided to consult no less an authority than Brahma, the creator of the universe.

After arriving at Brahma's abode, Kukudmi discovered that he had to wait since Brahma was busy. In fact, Brahma sent word that after his urgent work was over he will be free to see him. So along with his daughter, the king waited. True to his word, Brahma called him soon after and asked him the reason for his visit. Having given the background, Kukudmi stated the bottom line. Would the Great Brahma advise him as to whom amongst the existing aspirants should he choose as Revati's husband. On hearing the problem, Brahma laughed and said, "It is as well that you brought Revati with you. For while you waited here for about five minutes, the time has moved faster on your Earth. Perhaps a few million years have elapsed since you left to visit me. Naturally, all those young men you talked about have been dead and gone! So saying, Brahma gave him the name of Balaram (the older brother of Krishna) as the ideal son in law who will be around at the time Kukudmi would reach his palace.

This episode shows the radical idea that there is nothing absolute about time. Each observer carries his or her own timeline, and one needs to adjust the lines for the passage of time of different observers before comparing their measurement details. Einstein's relativity theory brought this fact to the attention of physicists. At first, it was not believed but experimental evidence was finally convincing.

In the modern era, we come across results similar to the Kukudmi Brahma episode. Astronomers report examples of massive stars which shrink under their own force of gravitation and eventually become black holes. As the shrinking star approaches the state of a black hole, its environment encounters a stronger and stronger effect of its gravitational attraction. One consequence of the same is that a clock near such a star goes slower and slower. One can imagine a thought experiment wherein two observers A and B set up a signalling protocol near a shrinking star as follows (Figure 13.5). A is an observer who is located in a fixed place observing the star from a distance while B is close to the surface of the star. B agrees to send time signals to A regularly, say, once every hour.

What happens in reality? A discovers that B's signals arrive slightly late, say initially one second late. While A is willing to condone an occasional delay, he discovers that it is progressively increasing. Indeed, as time progresses B's signals arrive after longer and longer intervals. Imagine that A is an applicant while B is a bureaucrat in a government office. A is expecting B's replies to his queries at regular intervals. But when they start getting more and more late, A will put it down to bureaucratic delay. But is B being late deliberately? No, his watch shows that he is sending signals as per the agreed protocol. The fact is that B's clock has slowed down compared to A's. *Indeed there may come a stage when no signal from B will ever reach A.* As Kukudmi and Revati found, the time passage near B had slowed down considerably as he continued to fall inwards. And using a modern version we could say that Brahma's abode is just outside the horizon of a black hole! The limit which is crossed by B in order that no signal from B will ever reach A is called the *event horizon* of the black hole.

How do we discover black holes? If a black hole cannot be seen, how do we know that there is a black hole in a certain part of the space? The answer is, a black hole is made of concentrated matter which has a large mass and hence a strong gravitational influence on the surroundings. For example, astronomers are familiar with binaries – that is, double-star systems. These are stars going around each other under the gravitational attraction of each other. If one of them becomes a black hole, we will see only its companion and from its circular motion deduce that it has a companion star. Why? Because Newton's laws of motion tell us that if the star were a single isolated one, it would have gone on in a straight line with uniform velocity. The fact that it is moving in a circular orbit suggests that it has a companion star which exerts a gravitational pull on it.

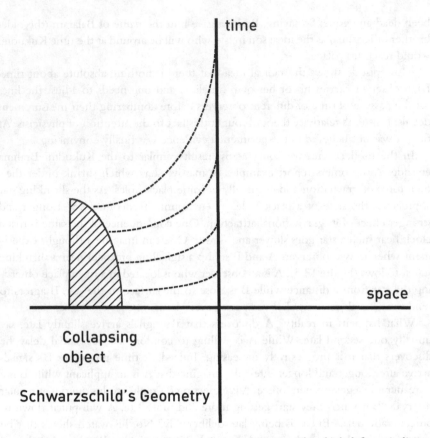

FIGURE 13.5 Collapsing body observed from a distance. The shaded form indicates a collapsing body. The dotted lines show how light signals travel from observer B on the surface of the collapsing body to a stationary observer A far from the body. Because of the increasing gravity of this body, the signals reach the receiver after longer and longer durations.

Many star pairs like these are known to be sources of X-rays, and these come because star pairs interact by generating a flow of plasma (not blood plasma but fluids of electrically charged particles) which flows from the visible star into the black hole. This in turn causes the production of X-rays. In the 1970s, space technology could place X-ray detectors in space, which led to the discovery of many such X-ray sources. From their studies, one could show that at least a few harbor black holes.

More recently, astronomers could detect a more bizarre system: a pair of black holes going around each other! How do we know that there are black holes out there? This technology is the recently deployed system of detectors of gravitational waves. A black hole pair churns up its gravitational environment and produces waves of gravity. The recent use of laser interferometers has succeeded in detecting such a black hole binary made of two black holes of around 35 and 29 solar masses, respectively. Note that in this case, neither of the two black holes will be visible to ordinary telescopes!

Clearly, black holes, the heavens of bureaucrats, exist in space and can be found given the benefits of technology.

Bending of light

With all its intricate mathematics, the theory began its life with the reputation of being very difficult to understand. A scientific theory becomes more and more acceptable if the expert community feels that it has been proved successful in several experimental or observational tests. There were very few tests that GR could offer. One of those, which was to play a crucial role in lending credibility to the theory, concerned the so-called bending of light by gravity. Although Isaac Newton had been non-committal during his life in arguing for or against there being a bending of light by the Sun's gravitational attraction, other scientists were keen to settle the issue. Indeed in 1801, using Newton's own ideas and laws of physics, Johann Soldner calculated that a light ray grazing the Sun's surface would be bent by an angle of 0.87 arc seconds. In practical measurements, this corresponds to such a tiny measure as about 4,000th part of a degree.

Before he finished putting touches on his theory, Einstein calculated the bending angle in the same case that Soldner had worked on. *And he got the same answer as Soldner.* Whereas the Newtonian prediction had been bogged down with several assumptions, Einstein's approach was straightforward, and it gave the same answer: at least so everybody thought back in the year 1911 when Einstein did the calculation.

It was known that the bending of light from a distant star can be tested if the light ray passes close to the Sun. A bend in the light track would lead to the shifting of the star's image by a very tiny amount. Thus, one needs to measure the positions of the star's image when its light grazes the Sun and when it does not. Evidently, the first part of this experiment appears impossible to do. How can an astronomer record the image of a star when the Sun is shining nearby? Clearly, this is impossible except on a special occasion which must be used – namely, the timing of the experiment during a total solar eclipse.

After Einstein's 1911 prediction, a German team of astronomers set out to test it by using the total solar eclipse of 1914. The onset of the First World War, however, vitiated that plan. Indeed, the Germans found themselves in the position of what the war defined as "enemy nationals." In the meantime, Einstein redid the calculation and realized that he had made a mistake, and the correct answer was 1.74 arc seconds – that is *twice the Newtonian value.* Looking back, we see how the circumstances spared Einstein an embarrassment. For if the German team had been allowed to carry out their observations, they would have come out with a result that contradicted Einstein's earlier (wrong) prediction.

Finally, for the next eclipse in 1919, two British expeditions were proposed under the leadership of the Cambridge professor Arthur Stanley Eddington. In those early days, Eddington was one of the few scientists who really understood what relativity was all about and who felt that such a marvellous theory should be tested by observations. However, one major difficulty arose. In the war years 1917–1918, there was conscription in force in Britain, and as a conscientious objector, Eddington was liable to be put in detention. Many scientists and public figures urged that this should not happen, as Eddington was engaged in planning an important experiment. Fortunately, the Draft Board accepted this argument!

The two locations chosen for the experiment were Principe in Spanish Guinea in West Africa and Sobral in Brazil. Eddington himself opted for the former and another astronomer Andrew Crommelin was put in charge of the latter. They took two specialist telescopes with objective glasses of 10 m diameter each. As an afterthought, the Sobral team also took a tiny 4-inch telescope as a 'backup.'

The actual observation in both places had its own moments of chaos and confusion. Not having allowed for effects of heat and readjustment in the telescope for change of latitude of observation, the Sobral team found that their main telescope was unusable and so they used the backup instrument. Though Eddington had chosen the Principe site because of better weather; in reality, there were thick clouds and heavy rain. Fortunately for him, in totality, the weather cleared enough for taking the crucial star photographs. And, finally, they needed to take photographs of the same stars when their rays did not pass close to the Sun. The actual bending of light could be estimated only from differences of two such measurements (with and without bending of light; Figure 13.6). However, an impending

FIGURE 13.6 Bending of lightThe distant star would not be seen because the Sun is on the way even in the eclipse. The bending of light by the Sun's gravity makes the star visible.

boat strike did not allow Eddington the extra time for taking the comparison photographs. These were taken in England later.

Nevertheless, as a pioneering experiment, its outcome was eagerly awaited by the scientific community. If the actual bending was close to 0.87 arc seconds, Newton would be vindicated, whereas double that value, 1.74 arc seconds, would favour Einstein. On 6 November 1919, at a crowded meeting of the Royal Society and the Royal Astronomical Society, Eddington announced the verdict: *Einstein wins.* The two sets of data after due analysis gave a value of 1.61 arc seconds with an estimated error of 0.4 arc seconds.

The rest is history. While it led to a greater popularity and level of acceptance for relativity, Einstein himself became an icon followed with interest and respect wherever he went.

But a second look at the 1919 experiment tells us why we need to attach a greater uncertainty to the conclusion drawn by Eddington and partners. There were several intrinsic sources of uncertainty which were not allowed for. Moreover, Eddington appeared to have taken out data points favourable to Newton, although he gave justification for doing so. As mentioned before a comparison photograph could not be taken on-site. Although later they took one in Oxford, the credibility of the result got compromised.

In fact, the really credible measurements of the effect come from radio and microwave observations which were carried out around 1975, some 60 years after the creation of the general theory of relativity. We did remember this landmark as we celebrated its centenary.

The lighter side of gravity

When the falling apple supposedly inspired Isaac Newton to discover the law of gravity, he did not imagine that he had opened a Pandora's box. For the journey from the falling apple has led to several strange consequences, like black holes. We will look at some by way of illustration.

The law of gravity as enunciated by Newton says that there is a force of attraction between any two bodies in the universe. The force is in direct proportion to the masses of these bodies and in inverse proportion to the distance separating them. To understand the full implications of this statement, we will compare it with another force which is familiar to us in daily life, *viz.* the elastic force which arises when we use a rubber band.

Tie two bodies to the ends of a rubber band and stretch it. The natural tendency of the rubber band is to contract to its original unstretched size. The two bodies that were tied to its ends will therefore feel a force of attraction because of the band's tendency to shrink. Is this not like Newton's gravitation, which also causes the two bodies to attract each other? This may appear to be the case, but in reality, there is a world of difference between the two cases. As we saw with the rubber band, it tends to contract when stretched, but this force of contraction disappears as soon as the band is allowed to shrink to its natural length. Thus, as its demands (to attain its natural length) are met, the force disappears. Not so with gravitation! If we yield to it and allow the two bodies to approach each other, *the force does not disappear. Rather, because of the inverse square law, it gets stronger.*

This feature makes gravitation stand apart from other forces in nature, as the elastic force disappears once its demands are met. The physicist Hermann Bondi called this tendency of gravitation 'dictatorial.' If you give in to the demands of a dictator, he makes further demands. As we will see, this dictatorial tendency leads to the formation of black holes.

Before we come to that aspect, let us examine another situation wherein gravitational force behaves strangely. This involves what scientists call a thought experiment – i.e., an imaginary experiment. Suppose you connect a hot body to a cold one with a heat-conducting wire. What will happen? Heat energy will flow from the hot body to a cold body with the result that the hot body will become cooler and the cold body hotter. This process will go on till both bodies have the same temperature. But now imagine that you are connecting two stars A and B of which A is considerably hotter than B. As before, heat will flow out of the hotter star A to B. Because A has lost energy, its internal equilibrium is upset. Normally in a star like the Sun, two internal forces are at play. The force of gravity tends to highlight the force of attraction between its different parts with the result that the star as a whole tends to contract. The star is, however, able to maintain a fixed size because the gravitational force is balanced by a thermal force arising from the fact that the star is made of hot plasma. In our thought experiment, this exact balance is upset. Since the star A has lost heat, it finds its interior deficient in strength, especially in balancing gravity. So it contracts and goes on doing so, and in the process, its internal temperature rises. This in turn raises the thermal (gravity opposing) forces and in the ensuing readjustment, the star's temperature rises.

What happens to star B? Recall that it receives energy from star A which results in the strengthening of its thermal forces. This results in an expansion of the second star. Because by expansion, hot gas and plasma cool down, the overall temperature of the colder star B will be lowered by our thought experiment.

So what is the bottom line? The experiment results in the hot star getting hotter and the cool star getting cooler! This counter-intuitive result arises because gravity is at play.

In reality, we may have such a situation with a somewhat different scenario when a star becomes a red giant, The interior of such a star has a hot core surrounded by a cooler mantle. The conditions of equilibrium have to accommodate this reality. So what happens? The core emits heat and contracts, but this process raises its temperature. The mantle receives that heat, which makes it expand. As an expansion of a gas or plasma leads to its cooling, the envelope cools down. Thus we have a large star which is cooler at its outer boundary but hotter at the centre of its core. Having grown large in size because of expansion, the star is called a giant. As its outer envelope is cooler the star has a reddish appearance. Hence the name "red giant."

The role of thermal pressures in maintaining stellar equilibrium cannot be overstated. Imagine our Sun suddenly losing all its pressures. With nothing to oppose its gravity, the whole mass of hot plasma will collapse, and the star would shrink to a point. How long will it take to shrink to a point? A mere 29 minutes! Not our Sun, but more massive stars may find themselves at such a stage when they have exhausted their nuclear fuel. With no resistance to their own gravity, such stars would undergo rapid gravitational collapse.

The collapse process leads to a stronger and stronger force of gravity near the surface of the star, so much so that eventually its strong gravity will pull back even light. That is when we say that the star has become a black hole. In a sense, a black hole is the outcome of unrestricted contraction induced by gravity.

Note

1 Perihelion is the point in the orbit closest to the Sun.

14

SECOND SHOCK

The quantum world

The hydrogen atom

A simple picture of the smallest of all atoms, the hydrogen atom will demonstrate why the classical picture of the small world is defective. The classical picture has the negatively charged electron moving around the positively charged proton. But this is not the end of the story! For the electron is constantly changing its direction of motion and as such, it is accelerated. (Acceleration arises from the rate of change of velocity. Velocity can change if the moving particle keeps the same speed but changes its direction. If the electron is in a circular orbit, it is accelerated.) And a rule established by Maxwell's equations is that an accelerated charge radiates. So the electron continually radiates, and as it loses energy through radiation, it will occupy a smaller and smaller orbit and will ultimately spiral inwards to the centre of the atom. The time scale for this to happen is of the order e^2/mc^3. This is around 10^{-24} seconds, which is too small compared to the lifetime of an atom. In short, the atom as originally envisaged does not exist.

In contrast to this classical picture, what is observed is that the electron does orbit around the proton, but its orbit does not shrink. Instead, the orbit is stationary and thus the atom has a stable structure. How does one understand the rationale for this odd behaviour?

Suppose a priori we assume that the electron has any possible path allowed with the proviso that every such path will have some chance of materializing. It is like dice tossing, with the understanding that there are six possible outcomes and each outcome has a one-sixth chance. This tells us that the earlier Newtonian rule with a definitive trajectory does not work, but a sort of probabilistic approach seems to be more successful. Using clues like this, one can formulate modified dynamical rules, which are known as *quantum mechanics*.

DOI: 10.4324/9781003203100-15

Just as relativity theory made it necessary to abandon the intuitively acceptable theory of Newtonian mechanics so was Newtonian mechanics made to play a subservient role to quantum mechanics. By this change of paradigm, one has replaced a clear-cut classical concept with a probabilistic one. In the following example, we see how this can be done.

Black body radiation

With the understanding of thermodynamics growing both experimentally and theoretically towards the end of the nineteenth century, the notion of black body radiation (Figure 14.1) began to generate fresh challenges. In simple terms, we may approximate the 'black body' by the oven that bakes cakes and bread. In such an oven, there are heating agents or radiators which start by creating a lot of heat. Ideally, no heat is allowed to escape outwards from the oven. At the same time, the thermostat arrangement controls the overall temperature inside the oven. So after an initial stage equilibrium is reached, how may we specify the final equilibrium stage of the gas in the oven? The heated system is called *black body* since, ideally, it does not allow any radiation to escape. Seen from the outside, such a body looks black.

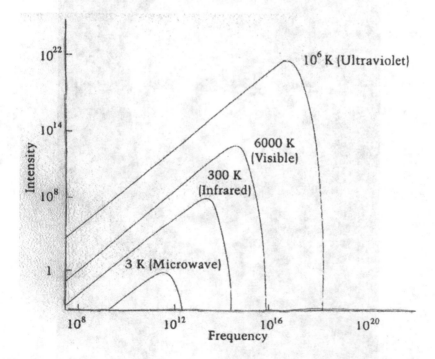

FIGURE 14.1   Black body radiation. The black body has a characteristic distribution of energy at different wavelengths. Some typical cases at different temperatures are shown.

If we start by calculating the thermal stage of equilibrium, we find that the Newtonian statistical mechanics fails to provide adequate details. The challenge was to describe the state of equilibrium of the gas which has been heated in the oven.

It was Max Planck who gave an adequate description by a pragmatic approach. Classical mechanics from Kelvin and Maxwell assumed radiation as continuous entities, whereas Planck assumed it to be a discrete collection of 'packets' of light. A typical packet is called a 'quantum' of radiation or a 'photon.' A photon has a frequency v and energy hv, where the constant h is called Planck's constant.

Figure 14.1 denotes the intensity of radiation plotted against temperature for black bodies of different temperatures. A typical curve shows the intensity of radiation rising with its frequency till it reaches a maximum value; thereafter, it goes down at a higher temperature. Thus provided we make assumptions as Planck did, we can understand the nature of black body radiation. The price paid for this understanding was the assumption of photon. In fact, this was the starting point of quantum theory, which we will briefly describe next.

PHOTO: Planck

The uncertainty principle

There is another way of bringing up for discussion the effect of measuring very small quantities. Normally searching for a macroscopic object, we may use a torch whose light helps us locate an average size quantity. Torchlight falling on the body tells us where it is. Since the body is not small, the torchlight does not disturb it. But the issue is not so simple if the body is not average size but is microscopic. For if we want to locate a microscopic quantity, the measurement may involve light, and the very process of measurement would involve a slight shift of the quantity. We would like to minimize the movement involved. Such calculations of results tell us that there is a minimum that might not be further reduced. This is stated as an inequality:

$$\Delta p \times \Delta q \geq h,$$

where h is called Planck's constant. Here p and q are the variables measured, and the errors are indicated by Δ. Thus Δp is the error in p and Δq the error in q.

In quantum theory, one identifies variables that are complementary. That means that if we set out to measure two such variables like p and q simultaneously, then we will discover that both cannot be measured with arbitrarily large accuracy; if we increase the accuracy of one, then that of the other will be reduced. This absolute limit on the accuracy of measurement is called the *uncertainty principle*.

While physicists were trying to express the conditions like the aforementioned mathematically, they discovered that the solution already existed. Thus, for example, they found that the complementary variables p and q can be shown to follow a non-commutative algebra:

$$pq - qp = ih / 2\pi.$$

Using this relation, we can easily show that $pq^2 - q^2p = 2q.\dfrac{ih}{2\pi}$.

A simple manipulation shows that from the previous relation, the variable p plays the role of derivative. Thus, we have for any function ψ of q *the following relation*

$$p \, \psi = -\{ih / 2\pi\} \, \partial\psi / \partial q.$$

So, in general, if $f(q)$ is any analytic function of q (e.g. having a power series expansion). we will get

$$f(q)q - qf(q) = f'(q).$$

Thus, it is possible to do all calculations for the interior of the atom using non-commutative algebra because the previous relation tells us how to use this algebra. The classical picture of the hydrogen atom is now changed to one where the continuous orbit is changed to a series of stationary orbits. The innermost orbit is the one most stable. Typically, an electron jumps to the lowest orbit in case it is not already there. The stationary orbits are obtained as eigenstates of the atom.

Indeed, it is one of the attractive features of quantum theory that algebraic derivations instead of calculus are used to bring quantitative results.

The double-slit experiment

Just as SR presented several counter-intuitive findings, quantum mechanics proposed a similar series of experiments at the microscopic scale of atoms and nuclei.

We will describe an experiment called the 'double-slit experiment' (Figure 14.2) in which a stream of electrons is released and directed to go through two parallel slits. What will happen? If we expect the electrons to behave like Newtonian streams of particles, we should see the two slits permit a discrete set of two streams. However, that does not happen! Just as a stream of X-rays gets diffracted, we see the electrons also 'diffracted,' and we see some electrons off the slit directions too. This type of experiment could be performed in many different ways. The flow of electrons can be modulated such that their observed 'passage' through the slits is totally unexpected.

This experiment was historically the early experiment which alerted the physicist that there are many interpretations which tell us that an electron is best described as a *wave function*, and its experimental behaviour is non-classical.

Another problem that emerged was that it was not possible to measure certain dynamical variables related to such particles. Intuitively, one could argue that if there are two complementary dynamical variables like position and motion of a particle, then there is a limit on the accuracy of their simultaneous measurement.

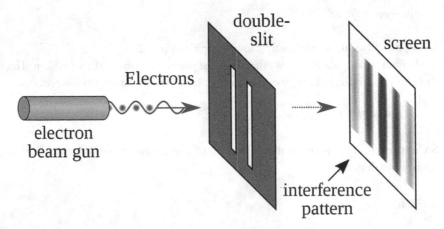

FIGURE 14.2 Double-slit experiment. A picture of the double-slit experiment as described in the text. The passage through the slits gives some unexpected results, showing a wave-particle duality unexpected in Newtonian physics.

If the particle has position q and momentum p, then the limits of their accuracy denoted by δq and δp are given by the product of their errors $\delta q \times \delta p$. This product is assumed to be given by Planck's constant h.

Wave-particle duality

When various experiments on subatomic particles like electrons, protons, etc., were performed, their outcomes were somewhat unexpected. The 'double-slit experiment' described earlier showed that the electrons are not behaving like classical charged particles. Rather, they behaved like waves.

The same duality applies to light. It behaves like a wave, as well as like a particle. The photoelectric effect is an effect that shows the particle nature of light (Figure 14.3).

While describing the Newtonian controversy about wave/particle nature of light, we have seen that the last word was not said on the controversy. The quantum mechanical perception resulted in a wave/particle dual nature of all subatomic particles. The 'photons' in Figure 14.3 are not the same as Newton's corpuscles. Their behaviour is more complex than Newton's ideas would suggest.

When radiation is enclosed in a closed volume, it reaches a state of equilibrium such that the behaviour of wave/particle duality plays a role here too. This 'trapped' black body radiation has a characteristic frequency distribution – a formula first suggested by Max Planck and later derived mathematically by S. N. Bose.

Hidden variables?

The uncertainty principle results in limits on measurements, and in some cases, there appears to be a reason ascribed to 'hidden variables.' These variables are supposed to be at a very small level so that they are not measurable. If normal variables require some knowledge of these hidden variables, then not knowing them may

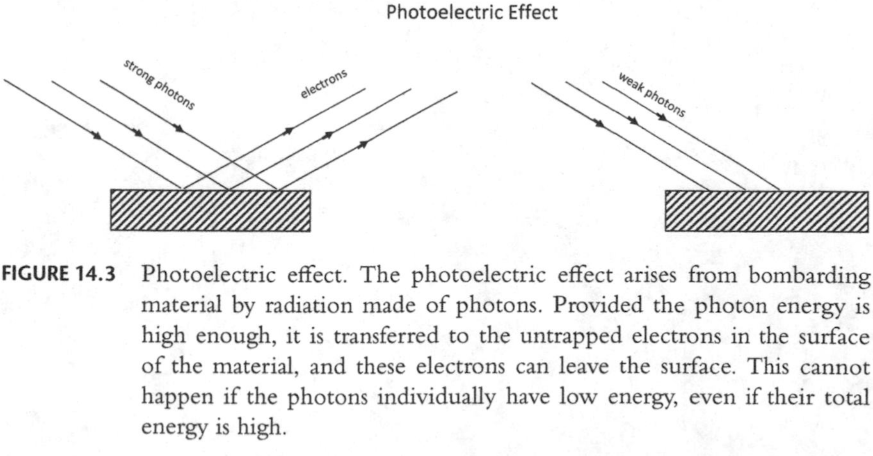

Photoelectric Effect

FIGURE 14.3   Photoelectric effect. The photoelectric effect arises from bombarding
material by radiation made of photons. Provided the photon energy is
high enough, it is transferred to the untrapped electrons in the surface
of the material, and these electrons can leave the surface. This cannot
happen if the photons individually have low energy, even if their total
energy is high.

lead to limits on the knowledge of normal variables. Several scientists, including Albert Einstein, Podolski, Nathan Rosen and others, proposed this solution to explain why there is a limit on knowing certain variables. It is worth noting that Einstein did not like the probabilistic nature of quantum mechanics.

However, several experiments designed to check the influence of hidden variables have so far not given any positive proof of their existence. Consider the following experiment. Suppose there is a source at a point S which emits two particles A and B in opposite directions. If the original source had zero spin, the emitted particles will have spins equal and opposite. An observer located in the path of particle B measures the spin of that particle. If the spin is positive, then the observer concludes that the spin of A must be negative. Notice that the information about A is known as soon as the measurement of B is done. Thus the special relativistic speed limit of the speed of light seems violated. One can avoid such a conclusion by invoking hidden variables.

This basic idea has been tested by various scientists like John Bell, John Klauser, Alan Aspect and others to see whether hidden variables are needed. But so far, there is no indication that hidden variables are playing a part.

We saw in the case of SR that several apparent violations were noted where kinematics of fast-moving (velocity close to that of light) particles were involved. Many physicists felt that there were manifest contradictions. A somewhat similar situation arose with the conclusions of quantum theory. As noted in the discussion between Einstein and Niels Bohr, even Einstein had problems with the interpretation of results of quantum theory. As Richard Feynman has commented, "Nobody understands quantum theory. But it works!"

15

THE CONQUEST OF SPACE

The early postwar years

From their primitive days, humankind has been impressed by the space around them and has wished to travel in it. Although humans could see that this required special powers, their imaginations generated 'gods' who had that ability. So in Indian mythological literature, such superhuman persons were presented prominently. In other civilizations, such mythical ideas are also found. It was much later when the Second World War was drawing to a close that the dream of moving around in space did not look bizarre or beyond practical human powers.

Towards the end of the Second World War, Nazi Germany had started using rockets and other guided missiles and the power behind them came out of chemical reactions. Calculation based on Newton's laws of motion and gravitation showed that if an object could be fired with a velocity exceeding approximately 11.8 km per second, it could escape from the earth's gravitational pull. That launch speed was mandatory and human efforts were directed towards generating a technology that made it possible.

After the Second World War was over, rival nations Russia and the United States got into an unofficial competition to achieve the aforementioned, and the first round was won by Russia. The announcement of the launch of Sputnik I by Russia on 4 October 1957 led to a stunned reaction by the rival USA. Realizing the need for an accelerated response, the USA established the National Aeronautics and Space Agency (NASA, as since known popularly). And amongst the quick actions undertaken by NASA was the upgradation of mathematics teaching in schools and colleges. The importance of mathematics in modern space technology was realized by the launch of Sputnik.

DOI: 10.4324/9781003203100-16

Man on the moon

Russia shortly scored another 'first.' On 12 April 1961, Russia sent an astronaut in a satellite, Vostok-I, round the earth, successfully bringing him back hale and hearty. This was Yuri Gagarin, and he had reached the height of 327 km during his orbit around Earth. Nettled by these 'defeats,' the US president announced a challenge for US scientists: to land an American astronaut on the moon and bring him back safely before the decade (1960–1970) was over. The US establishment rose to the occasion and the lunar landing took place on 20 July 1969. It was an American astronaut, Neil Armstrong, who uttered the famous line as he stepped on the Moon, "One small step for man, a giant leap for mankind." Although Russia did not send a man to the Moon, it used remote control photography to photograph the hitherto unseen other side of the Moon.

The Indian effort

It is against these mega efforts that one should view the foresight and dedication of Vikram Sarabhai in setting up the Indian Space Research Organization (ISRO) in 1969. ISRO had followed the more modest beginning of India's space programme in 1962, the Indian National Committee for Space Research. Earlier, the space programme was in the Department of Atomic Energy. Later, it had its own Department of Space.

PHOTO: Sarabhai

Remote sensing

Having been used to observing the universe from Earth, humans were not used to the experience of observing Earth from space. With satellite technology, humankind has been able to achieve that goal. Such an exercise is called *remote sensing*. One may wonder why do this. Don't we know our Earth, having lived on it for thousands of years? The answer is, *We did not have a complete global picture.*

Remote sensing provides information about Earth's surface at greatly desired accuracy. On a large scale, we may need to know forest cover, water reservoirs, mines of various sources, etc. On a small scale, we can locate streets, lanes, individual houses, etc. That is how we have a geopositioning system at work.

Telescopes

As we know, astronomers use optical and radio telescopes sited on Earth. For other wavelengths like microwaves, infrared, ultraviolet, etc., ground-based telescopes are no good, since all these radiations get absorbed by Earth's atmosphere as they come down. To have effective use of a telescope at these wavelengths, we need to place it at sufficient height so that no absorption has taken place before imaging. Thus all these telescopes are effectively satellites of Earth.

Looking at the advantage of placing a satellite/telescope in space, optical astronomers also wanted to place a space telescope at a suitable height. After several years of work, the Hubble Space Telescope saw the light of day in 1994. Although ground-based optical telescopes are common, the space version has several advantages. For example, the absorption of light from a distant source by Earth's atmosphere can be avoided by placing the telescope high up in space. Likewise, an image from a space telescope is considerably more steady than that from a ground-based telescope. This is because the atmosphere of Earth can be disturbed and would lead to unsteady light, which makes an unsteady image. In a space-based optical telescope, this unsteadiness is more or less absent, making the image very steady.

Communications satellites

Arthur C. Clark, an engineer by profession made a suggestion that proved to be enormously profitable. Consider, for example, a satellite going around in an equatorial orbit – that is, a satellite which moves at a height h, but stays above the equator. If Earth's radius is R, the dynamics of the orbit give the relation.

$$h = \left(\frac{R^2 g}{\omega^2} \right)^{\frac{1}{3}} - R,$$

PHOTO: Clark

where g is the acceleration due to gravity on Earth's surface, and ω is Earth's spin rate so that $2\pi/\omega$ corresponds to one day.

Thus, the previous relation tells us that at a height of approximately 35,800 km it is possible for a satellite to have the same angular velocity as any point on the equator. In short, for an observer on the equator, there will be a height at which a satellite will always stay overhead. Clark suggested that this location of a satellite will be ideal for its use for communications. All relay or broadcasting can be done by a suitably designed communications satellite launched to stay at its above optimum location.

Space travel

Following the lead given by Yuri Gagarin and Neil Armstrong, the next step might have been the launch of spaceships to the Moon, Mars, Venus and some satellites of Jupiter. But this did not happen because of various infrastructural issues to be sorted out. To enable the smooth transfer of astronauts from spaceship to Earth, the concept of the *space shuttle* had to be sorted out first. The shuttle is like a platform in space, and it can be docked against the parent spaceship, as well as other such platforms. The shuttle can also be used to do space repairs.

Two tragic accidents resulting in the death of all astronauts on board two spaceships caused the US programme to slow down. The Russian space programme also slowed down because of accidents, and a revival is now taking place. The Indian space programme has gone as far as spaceships to the Moon and to Mars. It is expected that the Indian space programme will shortly put a man in space.

Astrobiology

Following the developments of astrophysics and astrochemistry, the topic of astrobiology is slowly but surely coming into study. For example, we may want to know if bacteria and viruses are found in space. A provocative hypothesis by Fred Hoyle and Chandra Wickramasinghe stated that bacteria and viruses might be brought into the vicinity of Earth by comets in frozen form. When approaching the Sun, the comet warms up, and its tail is formed. The tail may carry the microorganisms, and if it brushes the upper layers of Earth's atmosphere, some of these may be transferred there and drop on Earth under gravity. Do such microorganisms exist?

ISRO has supported experiments to find the answer. In two such attempts in 2001 and 2005 a balloon-borne payload was sent up to height of 41 km and air samples were collected and brought down. The experiments consisted of biological study of the samples. Both experiments showed the presence of bacteria. In the 2005 study several bacteria, some not known on Earth were found. Further, experimentation is under preparation to obtain nuclear isotopic composition of such microorganisms to check if they could be extraterrestrial.

Clearly, we find an interesting outlet here for astrobiology and no doubt more will turn up as humankind becomes more and more interested in space!

16

WOMEN CONTRIBUTORS TO SCIENCE AND MATHEMATICS

Introduction

An examination of historical records would show that most countries failed to have any significant number of women scientists or women mathematicians. If we start counting names, we have Hypatia from Greece (see Chapter 1) or Leelavati from India (see Chapter 2), but the list would hardly progress further. In Cambridge in the days of Senior Wranglers, there was the singular case of Philippa Fawcett who got top marks in the Tripos examination but was not recognized as Senior Wrangler. Why? Because although women were permitted to appear in the Tripos, their candidature was not officially recognized. So Miss Fawcett, despite getting the maximum marks, was passed over in favour of the male candidate who had scored the highest amongst men but not as high as Fawcett. Likewise, women were permitted to take part in the Tripos examinations but were denied a degree, even if they otherwise qualified for it. It was only in 1948 that the rules changed, and Oxbridge degrees were available to women also. It is worth noting that the British set up new universities in Bombay, Calcutta and Madras in India in 1857, and these had no problem in issuing degrees to women.

The situation in Europe was no different! In Germany, for example, women were forbidden from attending lectures in universities. In Russia, female students were not admitted. In spite of such resistance, a few lady mathematicians or scientists did make it to the top level in their subjects. Here we describe some of those brave exceptions.

Sophia Kovaleskaya

The middle of three siblings, Sophia Kovaleskaya was presented with the remarkable world of mathematics in a rather unexpected mode! Her family in Moscow

DOI: 10.4324/9781003203100-17

decided to move to the countryside and as part of decorating the country house bought wallpaper of a specific kind to cover the rooms. However, they ran out of that paper with the children's room still not covered. As they did not want to go searching for more paper, they used the pages of a mathematical text for wallpaper. The kids thus had a wall full of pages containing mathematical formulae and equations to look at. Although the 11-year-old girl did not understand them, she was fascinated by them, and this went a long way towards influencing her to study 'this strange subject.'

Her family, seeing her interest in maths, appointed a tutor for her, and she progressed to A-level in need of professional university lectures. As a woman, she was not allowed to formally go to a Russian university. To study outside Russia, she was required to marry and have the permission of her husband. This she managed by having a dummy marriage and getting the husband to give the permission. In 1867, the couple left Russia, and in 1869, she got admitted to the University of Heidelberg in Germany where women were allowed to study. There she had the benefit of having teachers like Helmholtz, Kirchoff, Bunsen, etc. Later, she went to Berlin where she attended lectures by Weirstrass. But this was in secret since women were not permitted to attend these lectures.

Kovaleskaya did very good work to receive high acclaim and awards. But she was denied professorial positions at many universities, including those in Russia. In the United States, there is an Association of Women in Mathematics, which has a distinguished lecture named after her. She settled down in Sweden and changed her first name from Sophia to Sonya.

Marie Curie

Maria Salomea Sklodowska, alias Madame Curie, was born in Warsaw, Poland, in 1867, but because of her work, she is identified with France where she did her important work that won her the Nobel Prize in physics in 1903 jointly with her husband, Pierre Curie. In 1911, she won another Nobel Prize, this time in chemistry.

In her early life, she started her studies in Poland, but later she moved to Paris where she spent most of her career. There she and her elder sister Branislava both worked on heavy nuclei and their behaviour – how they break up into smaller nuclei and gamma rays. She called such break-up activity 'radioactivity.' This name has become standard now. She was the first woman professor at Paris University. As a mark of respect that her adopted country France felt towards her, her remains were interred in Pantheon in 1995. Her example opened the door for other talented women to get academic positions.

Not fully realizing the dangers of handling radioactive substances, she did not take adequate precautions and succumbed to aplastic ammonia in 1934. She is one of those very rare scientists who won the Nobel Prize twice.

PHOTO: Marie Curie

Amalie Noether

Born in 1882, Amalie Noether was a theoretical physicist, as well as an algebraist. Her skillful combination of mathematics and physics was extremely useful in the understanding of the symmetry of physical laws. Noether's theorem played a leading guiding role in understanding and describing the symmetry of any physical law under consideration.

Noether's father Max Noether was himself a mathematician, and Noether initially wanted to study the French and English languages but later felt that she liked mathematics more. She got a degree in mathematics. She started lecturing unofficially and without pay. Distinguished mathematicians like Hilbert and Klein wanted to invite her as a lecturer at Göttingen University but the prevailing antifeminist policy put a stop to that proposal. So she started lecturing under Hilbert's name. Her fame spread and several bright students came to work under her guidance. The algebraist van der Warden was one of these "Noether boys." His book on algebra actually was based on her work, which won her international reputation. Later, the spread of Nazi propaganda made it difficult and unsafe for her to continue working in Germany. She migrated to the United States. She was to work at the women's college Bryn Mawr as a professor. But in 1935, she died unexpectedly in surgery on an ovarian cyst.

She is well-known for her work on Noether's theorem, differential invariants, new garb for algebra, non-commutative algebras and hypercomplex numbers.

Rosalind Franklin

Born in 1920, Rosalind Franklin became famous for her work on molecular biology. Many scientists feel that she deserved a share of the Nobel Prize for her work on the structure of the DNA molecule. Because of her early death in 1958, she missed that recognition. She worked on coals under the auspices of the British Coal Utilization Research Association and got her doctorate in 1945. Later, she worked in London, first at King's College and later at Birkbeck College. Her DNA studies used the technique of X-ray diffraction.

It is known that Crick and Watson in their Nobel Prize work on DNA had used her X-ray images without her permission. Unfortunately, she was no more when the prize was announced in 1962.

We add a few examples from India. The episode of Vedic times often mentioned is that of Gargi and Yajnavalkya. We reproduce here the account given in *Brihadaranyaka Upanishada* by Vande Mataram Library Trust.

Dialogue between Gargi and Yajnavalkya

The following episode describes how a learned woman in Vedic times in India could challenge learned men as part of education.

Brihadaranyaka Upanishad 3:8

It was the court of King Janaka. Yajnavalkya received questions from all learned sages and seers assembled there, and he kept offering answers to all of them. Amongst them was a female sage Gargi, the daughter of Vachaknu. Addressing the assembly, she said, "Revered Brahmins, I shall ask Yajnavalkya two questions. If he is able to answer them, no one among you can ever defeat him. He will be the great expounder of the truth of Brahman."

Yajnavalkya said, "Ask, O Gargi."

Gargi said, "Yajnavalkya, that which they say is above heaven and below the earth, which is between heaven and earth as well, and which was, is, and shall be – tell me, in what is it woven, warp and woof?"

Yajnavalkya said, "That of which they say, O Gargi, that it is above heaven and below the earth, which is between heaven and earth as well, and which was, is, and shall be – that is woven, warp and woof, is the ether. Ether (*Akasha*) is the subtlest element. So subtle that it is often indistinguishable from Consciousness. Without it nothing can exist. Yet there is more."

Gargi said, "Thou hast answered my first question. I bow to thee, O Yajnavalkya. Be ready now to answer my second question."

Yajnavalkya said, "Ask, O Gargi."

Gargi said, "In whom is that ether woven, warp and woof?"

Yajnavalkya replied, "The seers, O Gargi, call him *Akshara* – the Immutable and Imperishable Reality. He is neither gross nor fine, neither short nor long, neither hot nor cold, neither light nor dark, neither of the nature of air, nor of the nature of ether. He is without relations. He is without taste or smell, without eyes, ears, speech, mind, vigour, breath, mouth. He is without measure; he is without inside or outside. He enjoys nothing; nothing enjoys him.

"At the command of that Reality, O Gargi, the sun and moon hold their courses; heaven and earth keep their positions; moments, hours, days and nights, fortnights and months, seasons and years – all follow their paths; rivers issuing from the snowy mountains flow on, some eastward, some westward, others in other directions.

"He, O Gargi, who in this world, without knowing this Reality, offers oblations, performs sacrifices, practices austerities, even though for many thousands of years, gains little: his offerings and practices are perishable. He, O Gargi, who departs this life without knowing the Imperishable, is pitiable. But he, O Gargi, who departs this life knowing this, is wise.

"This Reality, O Gargi, is unseen but is the seer, is unheard but is the hearer, is unthinkable but is the thinker, is unknown but is the knower. There is no seer but he, there is no hearer but he, there is no thinker but he, there is no knower but he. In Akshara, verily, O Gargi, the ether is woven, warp and woof."

Hearing these words from Yajnavalkya, Gargi again looked at the assembled Brahmins and said, "Revered Brahmins, well may you feel blest if you get off with bowing before him! No one will defeat Yajnavalkya, expounder of the truth of Brahman."

An Initiative by Vande Mataram Library Trust, Gurugram & Sri Aurobindo Foundation for Indian Culture Sri Aurobindo Society, Puducherry.

This or a similar account elsewhere shows that Gargi was an exceptionally well-read woman who could argue on philosophical points with the most accomplished male philosophers. Another lady whose name appears in Indian annals of history was Bharati, the wife of Mandan Mishra, a leading philosopher who had (unsuccessfully) participated in a public debate on philosophical issues with the original Shankaracharya. When Shankara defeated her husband, Bharati continued the debate in her own right, making Shankara request time to find answers to her questions.

Finally, if we go by the legend, Leelavati, the daughter of Indian mathematician and astronomer Bhaskara, is believed to have acquired a good knowledge of these subjects. Unfortunately, as we have noted earlier, the paucity of written material makes it impossible to probe into details of Gargi, Leelavati and Bharati.

These ladies attained their proficiency in maths or science despite the many problems they faced. A common understanding is that a woman is a housewife first and anything else afterwards, thus making womanhood a handicap to start with. But that is only one reason. As Margaret Burbidge, a leading astronomer with many distinctions found, in the early days, she was denied observing permission to use the 5 m telescope on Palomar Mountain. The reason? Because the facilities at the telescope site did not have a ladies toilet!

PHOTOS: Margaret Burbige; Kamala Sohoni

Take the case of Kamala Sohoni, a woman scientist born in 1912, who was denied admission to the lab of the Nobel laureate Sir C. V. Raman, in the Indian Institute of Science at Bangalore. The reason given by Raman was that women are incapable of research. But she argued against the reason given and managed to get admitted on one-year probation. But because of her good performance, she was given a year-by-year appointment. Having got an M.Sc. in biochemistry there, and she went to Cambridge where she got a Ph.D. after three years of research. Although she had a successful academic career, she kept encountering opposition.

EPILOGUE

The continuing story of mathematics and science

Futurology

This brings us to the end of what we set out to do: to bring out the way mathematics and science grew side by side right from primitive conditions in which the caveman learnt to survive and improve his conditions to the rather sophisticated living conditions his descendants enjoy today.

It is likely that having reached the present state, the question could be raised, What are the prognostications for tomorrow? Such an exercise in futurology helps in planning strategies for the future, but it invariably fails in its predictions. Any achievements predicted today are most likely to be realized well before their predicted date. So we refrain from any predictions about specific aspects but stick to qualitative statements.

Post-DNA in the twenty-first century

The discovery of the DNA structure essentially opened new pathways to the future of biology. So much so that the twenty-first century is often called the age of biology just as the twentieth century is called the age of physics. There are several ways in which development in biology could take place. A few objectives are listed next:

Food problem solved with a dramatic rise in food production
Environment and pollution strictly controlled
New drugs and new technology-enabled methods of treatment
An understanding of how the brain functions
A global solution to the problem of garbage disposal
A genetic record of each individual which can be accessed in an emergency

DOI: 10.4324/9781003203100-18

These are a few from a global menu, and one hopes that future research will deliver the keys to the items on the menu! As an example see the following:

The new technique, revealed by the University of Maryland School of Medicine in Baltimore, is known as Emergency Preservation and Resuscitation (EPR) and could provide doctors with more time to save emergency patients with life-threatening injuries. The new EPR technique, however, cools the body to as low as 10°C (50°F). Trials are being conducted by the university on patients with acute trauma – such as a gunshot or stab wound. To be a candidate for the treatment, their heart must have stopped beating (cardiac arrest) and they must have lost more than half their blood.

The physics menu

We are still a long way away from the ideal of the theory of everything. We have a grand unified theory which ironically unifies all known basic interactions with the exception of gravitation. How to expand the network to include gravitation? Neither Newton's law of gravitation nor Einstein's theory of relativity shows how to do this. The difficulty lies in our understanding of 'quantum gravity.' In the case of quantizing electromagnetic theory, there were many experiments around which could act as checks on theoretical models. The expected effects of quantum gravity are so small that it seems unlikely that a check on the prediction of a theoretical model could be made.

The previous experience has shown that theories that are well established yield useful technological applications. Information technology, for example, has benefited from such applications. The laptop, mobile (cell) phone, GPS, etc., are examples of this development. The remarkable progress of nanotechnology holds out a very productive technological vista.

Physics is expected to solve two outstanding problems: "How to achieve controlled thermonuclear fusion," and "how to use space technology to harness solar energy." Either will help resolve the outstanding energy problem.

Mathematical gifts

If we ask mathematicians whether anything of importance will occur in their field and whether any use can be made of it, we may get the respective answers of "yes" and "no." We have seen how new branches of mathematics have sprung up from time to time. The process may continue in the future also. But perhaps a greater stimulus comes from challenge problems. The seven bridges of Königsberg, the four-colour problem, Fermat's last theorem, the Riemann hypothesis, etc., had posed challenges to human intellect. There are the Hilbert problems which fall in the same class. These problems, 23 in number, were posed by the mathematician Hilbert at the beginning of the twentieth century. Whenever one of these problems is solved, it makes world news.

Fermat's last theorem was one of Hilbert's problems. There was a handsome reward for its solution. As we saw (Chapter 5), Fermat himself claimed to have solved it, but his marginal note in the book he was reading stated that because of lack of space, he could not give his proof there. The problem was solved by Andrew Wiles in 1994. The details are too complicated to be given here.

The solution to the four-colour problem raised a new kind of issue. In a simplified form, we can think of a couple of dice tossed around. As usual, when they fall, the face-up numbers are counted. If this experiment is done a large number of times, which sum will feature most frequently? The answer is 7. This can be worked out with a simple combinatorial calculation. In the case of the solution offered by Appel and Haaken, there were a large number of cases to be examined, and human counting efforts were not adequate. So these authors used computers to do the counting needed. Although the method was logically sound, some mathematicians would not accept the proof. There may be other occasions where the computer is indispensable.

Here is a problem that may require a computer: "What is the number of primes less than 10^{20}?"

Visiting aliens?

The present technology may not be able to provide for visiting planets around nearby stars. A simple calculation illustrates the issue. The Moon is approximately 1.25 light seconds away. (A quick reminder: distance travelled by light in one second is approximately 300,000 km, the distance known as one light second.) Our present technology takes about 50 hours to cover that distance. Alpha Centauri, the nearest star for us is 4.25 light years away. Assuming it has a planet, how long will our spaceship take to cover it? The answer will run to several hundred thousand years! An astronaut leaving today on this ship is not going to outlast the return trip. So what one could try is to deep-freeze the person and hope that by slowing down the metabolism of the astronaut we may hope that he or she will survive the trip. If we put together all the 'ifs,' it makes a highly dangerous trip. And even in the case of success – that is, a safe return of the astronaut – by the time he or she returns, several hundred thousand years will have elapsed! Thus the astronaut will return to an alien environment. Also, those who sent the astronaut on this trip will be no more.

In short, such a trip will be an engineering and health hazard, as well as bring results of no relevance to the planners of the expedition. So one expects that such facts will deter adventures of this nature, limiting the applications of space technology to visits to nearby planets of the Sun. More realistic programmes may include habitation on the Moon and Mars and possibly some moons of Jupiter.

Finally, we may not visit extraterrestrials but might receive visits from them if they are much more advanced than we are. Indeed, they may view our astronauts as primitive species just as we view the cavemen of the bygone past!

Usually, a book ends with the completion of its theme. This one will not have a last word. Rather, more inputs are expected in the future. We will mention a few.

We have referred to the 23 difficult problems set by the famous mathematician David Hilbert. Whenever one of them is solved, it makes 'world news' amongst mathematicians. As of 2012, three were unsolved, three were too vague to be solved and six were partially solved. The problem of proving the Riemann hypothesis is considered the most important of the lot.

Additionally, after Hilbert's death, amongst his papers was found one more problem, which is considered the 24th problem.

But so far as mathematics and science are concerned, the pattern of work has been set for the foreseeable future, viz. the mathematician works solo while the scientist works in a team. For example, the paper announcing the first detection of gravity waves had nearly a thousand authors! Also, in contrast to mathematics, scientific research costs much more, especially if it requires highly sophisticated instrumentation.

The unwitting cooperation between scientists and mathematicians has been an attractive aspect of this development and is expected to continue being so. However, the high cost of modern scientific research may have an adverse effect on the valuable character of objectivity, as seen from the following episode.

While admiring the scientific progress, the following story told by Nobel laureate S. Chandrasekhar illustrates this point. The proposal to have a 200-in telescope on Palomar Mountain in Southern California was approved, and a press conference was called to meet two celebrated scientists: Edwin Hubble (celebrated for the discovery of the expanding universe) and Arthur Stanley Eddington (a pioneer in the theoretical studies of stars). One question the press asked was, "What do you expect to see with this instrument?"

The answer they gave was, "If we knew the answer, there was no need to have this telescope."

The modern attitude is different. First, a group of scientists has a theory which they wish to make popular. They approach the funding source with a fully prepared report on what they expect to prove. A peer review then takes place, and if the proposal agrees with the beliefs of the peers, there is a good chance of getting the funds. However, if the peers do not take kindly to the proposed viewpoint, the proposal will not be funded. This system, although regarded as 'fair,' does not encourage radically new ideas. It would not be wrong to say that if the heliocentric theory were proposed by Copernicus, the present review system would not have given support for it.

Nevertheless, we have to hope that an open mind will eventually prevail, although the time taken to get a new viewpoint accepted will take a longer time.

It was through keeping an open mind that our caveman progressed to become a space-riding astronaut. Let us hope that the future will reveal more of its secrets to the scientist and mathematicians will find more avenues to satisfy intellectual creativity.

APPENDIX A

In a lighter vein: some anecdotes about scientists and mathematicians

Although scientists and mathematicians have a reputation of being serious about what they do, there are occasions when they generate humour. Here are some examples.

1 Goa and gravitation

The following tale comes from the late C. V. Vishveshwara from Bangalore.

Many were the reasons for selecting Goa as the venue for a conference on gravitation. For instance, the natural beauty of the land with the Emerald Sea. The azure sky and the vast stretches of golden sand. The warm hospitality and the open friendliness of the people. A fascinating culture in which the East and the West have mingled together.

In addition to all these, there was a historical reason as well. Legend has it that a sage belonging to this region discovered the universal law of gravitation some 300 years before Isaac Newton. It so happens that there are hardly any apple trees in Goa, but one can find coconut groves all around. Consequently, the discovery of the law of gravitation by our sage was occasioned by the fatal fall of a coconut; the world remained ignorant of his finding. This was indeed the first case in the unrecorded history of perishing without publishing.

2 The sword of Tipu Sultan

Tipu Sultan was a king with the reputation that he honoured and rewarded arts and skills of various kinds. Naturally, his court always had some visitors wanting to show their skills.

One day, a claimant came with a very unusual skill. He was presented to Tipu and explained to the king what he could do. "I can throw this grain of rice through a one-millimetre hole in this metal plate from a distance of three metres," he claimed. Tipu looked impressed and asked the claimant to demonstrate the skill.

The rice thrower set himself up with a cupful of rice grains, 3 m from the metal plate, which was resting in a vertical plane. Then with a bow to the king, he started throwing the rice grains from the cup he was holding. A 'referee' observing near the metal plate declared that the rice thrower had achieved 100 percent success.

"Good!" said Tipu and called for his silver sword. The claimant and the courtiers present thought that the king was going to present the sword as a reward. But Tipu had other ideas when he said to his vizier to behead the claimant with the sword.

"Why this punishment, sire?" asked the claimant on behalf of all present.

Tipu replied, "You have wasted your precious life acquiring a thoroughly useless skill. The punishment is for this reason. I am doing you the honour for your skill by electing to behead you with this silver sword."

Pure mathematicians beware!

3 Amalie Noether and public baths

Amalie Noether was a German mathematician around the early twentieth century. Although her research work in mathematics was amongst the best of those days, she had to face antifeminist objections on numerous occasions.

The famous mathematician David Hilbert was very much impressed by Noether's work and invited her to Göttingen. Although there was a position vacant in the mathematics department of the Göttingen University, Hilbert's attempts to get Emmy Noether in that position met with stiff resistance by the philosophers at the university.

Fed up as he was by opposition to appoint the best candidate just because she was a woman, Hilbert expressed his disgust saying, "We are filling an academic position at a university, we are not discussing admission to public baths." But his urges were no use, as the antifeminist lobby was too strong.

4 A matter of statistics

The Nobel laureate Paul Dirac was reputed to be a man of few words. Once, on a visit to India, he and his wife were to land in Calcutta (now renamed Kolkata). Appropriately, Professor S. N. Bose (of Bose statistics fame) was deputed to receive them at the airport. Bose had taken with him the roomy Ambassador car, as well as two students to help out with luggage, etc. In the old days, a car used to have a continuous front seat on which three people could sit.

The plane duly landed, the Diracs came out and Bose conducted them to the car. He seated the couple on the back seat while he himself, the students and the driver squeezed themselves onto the front seat. Seeing the overcrowding on the front seat, Mrs Dirac said, "Why can't Professor Bose come over to the back seat . . . there is plenty of space here."

Bose was too embarrassed to say anything, but Dirac softly supplied the answer: "It is a matter of statistics, my dear!"

(We may add that in quantum theory there are two ways of filling empty spaces with particles. The Dirac method limits each space to at most one particle, whereas the Bose method has no such limitation. These methods are called Fermi-Dirac statistics and Bose-Einstein statistics.)

5 Who was testing whom?

This episode dates back to the nineteenth century, at a time when there was stiff competition for the top position in the Mathematical Tripos examination. Those who got first class were called 'Wrangler' and the top amongst them was called 'Senior Wrangler.' Two students named Parkinson and Thomson were considered the top two in this race for Senior Wranglership. In the end, it was Parkinson who stood first, closely followed by Thomson. But the head examiner had a nagging problem with regard to the performance of these top students.

The situation was as follows. In those days, the Tripos papers contained questions of variable standards. The simpler questions carried fewer marks and were intended to be solved by the 'weak' students like those who spent a lot of time in sports or other extracurricular pursuits. Such students, therefore, got enough marks to pass the examination and qualify for a pass degree. On the other hand, there were very difficult questions too which attracted the scholars who hoped for a high position in the wranglers' list and an honours degree. Parkinson and Thomson were naturally expected to solve the difficult questions. Which they had done.

However, there was an exceptionally difficult question which only these two had managed to do correctly. None of the other scholars had managed it. The problem that bothered the head examiner was that both had solved the question in exactly the same way. This had led him to suspect whether one student had copied the other. To settle this question he called the two students to quiz them further. First, he asked Parkinson, who was the Senior Wrangler, how he managed to complete the difficult question? He replied that besides the maths texts he had been reading research journals, and he had come across an anonymous article which did precisely the same problem that was now under discussion. As he remembered the steps towards the solution, he reproduced them when he saw the same topic appear as a Tripos question.

The examiner was satisfied with this reply, and he complimented the scholar for his practice of reading research journals while still an undergraduate. He then called Thomson, feeling sure that he would not come up with the same explanation. To anticipate him, he said somewhat belligerently: "Parkinson has just explained that he could solve that difficult question because he reads research journals and had happened to read there an anonymous article which covered the same area as the question. Don't tell me that you saw it in the same research paper."

"No, sir!" replied Thomson. "As a matter of fact, I wrote that paper."

This anecdote indicates the high level of the Tripos examination and how it brought up exceptionally bright students. The scholar Thomson later became a well-known physicist who worked on electricity and magnetism and thermodynamics and who was honoured with the title Lord Kelvin.

6 An 'easy' problem for the sportsman

A later example describes the consequence of attempting an apparently easy question. The story relates to Karl Pearson who in his later life became a pioneer in the field of mathematical statistics.

While appearing for the final examination of the Cambridge Mathematical Tripos, Karl Pearson noted that the desk ahead of him was assigned to a friend of his who had, however, no pretence to being a scholar, least of all a mathematician. He was a sportsman and hoped to do a few simple questions to get a pass degree.

Located behind him, Karl could very well see how his friend was faring. After about half an hour, Karl found that the sportsman had ticked off a question. If this fellow can do that question within half an hour, the question must be very easy. At least so concluded Karl, and out of curiosity, he took a look at that question. Contrary to his expectation, he found the question not so easy. Indeed, Karl felt that if the sportsman had done the question, he must have chanced upon some trick that made the question simple. Following this reasoning, he tried all sorts of mathematical tricks that he knew but to no avail.

Thus Karl got frustrated and felt it a challenge to solve that apparently simple question. He tried all methods that he knew to get the solution. At last, he succeeded but at a price. The attempts at a solution had taken up all the available time for the paper. So, he ruefully felt that his obstinacy had led him to spend almost all his examination time on a simple question. At this rate, he would be sure to fail the examination.

As he emerged from the examination hall, considerably depressed, he sighted his sportsman friend also emerging, apparently satisfied with his performance. Karl stopped him and asked the question that had been bothering him. How had he managed to solve that particular question? Could he tell Karl the trick he had employed?

"What trick? Indeed you are mistaken in concluding that I solved that question," replied the sportsman.

"Look at the question paper . . . you have ticked that question." Karl triumphantly showed his friend the tick mark.

The sportsman laughed and said, "Karl, I could not do a single question in that blessed paper. I was merely doodling to pass the time. The tick mark was part of my artwork!"

Later Karl discovered that he had been the sole student to correctly solve what was the most difficult question on the paper. His efforts were not wasted! When results came out, Karl was the third Wrangler.

7 Who stood first?

James Clark Maxwell is well-known for his work on unifying electricity and magnetism through a set of equations. When he was appearing for the Mathematical Tripos, the highest mathematical examination in Cambridge, he was confident of standing first. He was aware of two or three competitors, one of whom would stand second in the examination.

The result of the mathematics examination was read out by the chairman of examiners from the balcony of the Senate House at the stroke of nine in the morning. Since Maxwell did not want to get up early for the reading ceremony and was sure of his own result, he sent his servant to the Senate House with the instruction to get the name of the student standing second.

When the servant came back from the Senate House, Maxwell eagerly asked him, "Well, tell me who stood second?"

"You, sir!" replied the servant.

8 A friend of numbers

The Indian genius Ramanujan was hospitalized in London where his friend and mentor Hardy called to see him. While chatting, Hardy mentioned that he had come in a taxicab with the number 1729. "Not a very interesting number!" commented Hardy.

At this Ramanujan shook his head and remarked, "On the contrary, it is a very interesting number. . . . It is the first such number that can be expressed as the sum of two cubes in two different ways."

Hardy did some mental arithmetic and found that the young patient was right! One can write

$$1729 = 1^3 + 12^3 = 9^3 + 10^3.$$

This was the occasion when Hardy is reputed to have remarked that all numbers are Ramanujan's friends.

9 An Indian protocol?

The orthodox Indian mathematician Srinivas Ramanujan took some time to settle down in the alien atmosphere at Cambridge England. Being a strict vegetarian, he was reduced to cooking for himself. Nevertheless, when he felt settled, he took the bold step of inviting some friends to lunch.

The invitees duly came, and the lunch commenced. At some stage, Ramanujan offered to his guests a second helping of 'rasam,' the south Indian soup which is known to be somewhat peppery. The guests politely declined the extra helping offered.

Thereupon Ramanujan got up and walked out of the dining room. The guests assumed that he might have gone to fetch something. But time passed and still no sign of the host. After a decent wait, the guests decided to leave. They were of course mystified by the strange behaviour of the host.

What had happened was that the fact that none of the guests took a second helping hurt Ramanujan, who felt that his meal was not liked by the guests. Rather than be with them, he wanted to be somewhere else. The only such place he knew was in Oxford. So he went to the railway station and got a ticket for Oxford. He returned the next day by which time his guests had dispersed.

Evidently, there was a gap in understanding on both sides as to the protocol to be followed at an orthodox Indian lunch party.

10 The turban of C. V. Raman

Sir C. V. Raman, the Nobel laureate in physics from India, was being feted on the occasion of his 80th birthday. At a dinner where he was the chief guest, there were speeches, mostly by his former students and colleagues praising him. In the end, Raman got up to reply.

He began by thanking the hosts and then said that he was asked why he wears a turban. The reason he gave was the following. He said that the turban was like a bandage for his head: it prevented the head from getting swollen after hearing speeches offering him fulsome praises!

11 The elephant and the fly

I had an invitation from the Nobel laureate Subrahmanyan Chandrasekhar to report on the first Texas Conference held in Dallas in 1963. My report was scheduled in the afternoon slot at the Yerkes Observatory.

Before the review talk, Chandra took me aside and clarified one point. He said that at these seminars he tends to be very critical and asks very pointed questions. He told me not to take these interruptions as personal attacks. He gave an example. When astronomer Raymond Lyttleton gave a seminar, Chandra frequently interrupted, prefacing his doubts each time with the phrase, "I do not know how you got here" till the speaker was irritated and said, "The elephant may crush the fly saying, 'I don't know how you got there.'"

12 Laplace and Napoleon

Pierre Simone de Laplace had followed Newton's work on the mechanics of the solar system. He assigned to himself the problem of planets, satellites, asteroids, etc., all moving under the law of gravitation. Finally, he published this mammoth work under a five-volume book named *Mecanique Celeste*.

This book went a long way in giving credibility to Newtonian mechanics. When Laplace presented a five-volume set to Emperor Napoleon, the latter had a cursory look at the work and noticed one point, which he mentioned to the author: "I do not see any mention of God anywhere?"

To this Laplace replied: "Sire, I did not need that hypothesis anywhere."

13 Which is the first one-way street?

The Royal Institution (RI) of Great Britain was set up in the nineteenth century by Michael Faraday. Faraday not only was a great scientist who invented the electric dynamo and the electric motor but also spent time in public outreach. His public lectures and lecture demonstrations in the RI were crowd pullers and that tradition is still going strong two centuries later.

When these lectures and demonstrations were launched in Victorian England, the crowds became so large that the road on which the RI stands, Albemarle Street, had to be made a one-way street.

That happens to be the first one-way street on record!

14 Feynman in action

I was attending a small but select meeting on the nature of time, which included several distinguished scientists and philosophers of science. Amongst the participants were John Wheeler, Fred Hoyle, Philip Morrison, Dennis Sciama, Richard Feynman, S. Chandrasekhar and Adolf Grunbaum.

The organizer, Tommy Gold, informed us in the beginning that the sessions were in the round-table format, with each participant given a chance to make a presentation. All the proceedings would be tape-recorded and transcribed for publication.

While there was general agreement on this procedure, one participant (let us call him X) was against it. "If you are recording everything then count me out. . . . I will not participate."

Gold asked X to say why he was so negative. He said that the subject was such that he would be tempted to make any nonsensical remarks: which he would hesitate from making if he knew that these will be published. Although others said that even nonsense from him would be worth considering, X was adamant. Finally, as a compromise, it was decided that the participant will not appear in the proceedings as himself but as "Mr X."

This was followed; although from the proceedings, it is not difficult to identify Mr X.

15 Feynman's nonsense language

The International Conference in General Relativity and Gravitation was just over in Jablonna, near Warsaw. The year was 1962, and I was one of the student participants at the conference. One of the post-conference tours arranged by hosts took us to the hill resort of Zakopane. En route, our bus stopped for a coffee break and all of us were told to be back on the bus within 20 minutes. So we all did, except the driver, who was probably enjoying a liquid diet.

As all of us waited till the driver returned, one member of the group took control in his hands. He got down from the stationary bus and started talking loudly but in a language none of us understood. But the tone of the speech showed that the speaker was in some crisis. That produced the result, as the driver came running and started the bus.

The speaker of the strange language was none other than Richard Feynman. He explained that this was a nonsense language he had invented, and it always worked. Since nobody could understand what was said, the listeners thought that the speaker was calling for urgent action.

16 Collaboration? What has maths to offer?

A rider in Texas entered a small town after a long ride. He saw a shop advertising a laundry. "Your clothes washed while you wait," it said on a big sign. He went to the shop and took off his jacket and trousers, requesting the shop attendant to clean them.

"But we don't have a laundry here, sir," said the man behind the counter.

The visitor pointed to the sign and asked, "What about this claim?"

The shopkeeper laughed and clarified, "That is one of the many signs we paint." It was a sign painter's shop and not much use to our visitor!

Often, physicists are in need of a mathematical framework, but what mathematicians may offer will not be of any practical use!

17 The use of a barometer

In an oral test in physics, a student was asked how he would measure the height of a building with a barometer, provided access to the roof of the building was possible. He was told that he could use other supporting equipment. After a few minutes, he came with his solution.

"I would drop the barometer from the roof to the ground and with a stopwatch measure its time, t, of fall to the ground. Since I know g, the free-fall acceleration due to gravity. I have the answer as $1/2\, gt^2$."

Although the barometer was used, this was not what the examiner expected. So he asked the student to try again. He came up with the following answer:

"Tie the barometer to a long string and lower it from the roof till it touches the ground. Measure the length of the string."

"Try again," said the examiner.

The student replied, "I would tie the barometer to the string as in the previous experiment and make it oscillate with the lowest point just touching the ground. From the period of oscillation, I can calculate the length of the string and hence the height."

"Don't you know the answer that pressure drop with height can be measured by the barometer? There is a standard formula for it," said the exasperated examiner.

"I know that, sir, but it is more fun finding alternative ways of arriving at the answer," replied the student.

18 The role of the collaborator

The following story was told by a young scientist who was heavily criticized for presenting a paper using highly esoteric concepts. By way of retaliation, he told the audience the following story.

In a forest containing a wide variety of animals, a rabbit was accosted by a wolf. "Be prepared for I am going to eat you now," said the wolf.

"Don't do that, please. I am in the midst of very basic research," said the rabbit.

"I don't believe you," said the wolf.

"Then first meet my collaborator."

Thinking that the collaborator may be another rabbit and that two together would make a big feast, the wolf agreed to accompany the rabbit to his laboratory. This turned out to be a cave. The wolf entered the cave with the rabbit. However, he has not been seen since then.

The same development took place with the tiger, the bear and some other beasts, all ferocious but none came back since entering the cave. At last, the more adventurous amongst them decided to have a quick but silent peep into the cave to see the rabbit's collaborator. When they saw him, they understood why none came out.

For the collaborator was a gigantic lion.

The young speaker at the conference then confirmed that his collaborator was a Nobel laureate!

19 How modern science works

Dr Bambang Hidayat, a senior astronomer from Indonesia, has given the following examples of developments in modern science.

Patrick's Theorem (more advanced than Murphy's law): If the experiment works, you must be using the wrong equipment.

Compensation Corollary: Any experiment is considered successful if no more than half the data must be discarded as discrepant to obtain agreement with the favoured theory.

Axion of Credibility: Tell a nan there are 300 billion stars in the universe, and he will believe you. Tell him that the bench ahead has wet paint on it, and he will touch it to make sure.

Error Principle: To err is human but to really foul things up requires the computer.

Sameer's Algorithm: If you can't understand it, it is intuitively obvious.

Determinism: The only perfect science is hindsight.

20 Expert advice

Sir Harold Jeffreys, plumian professor of astronomy and experimental philosophy at Cambridge, was a distinguished theoretician and geophysicist. He was also acting as a consultant to an oil company. He sat through a meeting silent and puffing at his cigarette. Halfway through, the discussion stopped, and the chairman sought the expert's opinion.

"I think it is time for coffee," said Sir Harold.

After a coffee break, there was another discussion, which Sir Harold silently listened to until they asked his opinion.

The opinion came promptly: "I think it is time for lunch."

A similar development occurred at teatime.

Finally, before winding down the discussion, Jeffreys was asked, "What do you make of it all, Sir Harold?"

The great oracle replied, "I am glad it is your problem and not mine!"

21 No language barrier?

Astronomer and science popularizer Patrick Moore once narrated a personal experience. His regular programme called *Sky at Night* ran for several decades on BBC TV. He was very good at talks and interviews. One day, he was called by the BBC. A Russian astronomer was to be interviewed. Time was short, and Moore was requested to come straight to the studio. He was told that the Russian understood English but did not speak it. The BBC person said, "If you ask a question in English, he will reply in Russian, which you will translate into English for our audience."

"Hey!" cried Moore. "But I do not understand Russian!"

"That is your problem," replied the BBC man as he hung up.

In the actual interview, Moore guessed what the Russian had said and relayed it with great confidence. However, he was not sure how good his guessed reply was, and he was afraid that somebody in the audience who knew Russian might complain. So he was tensely waiting for time to pass.

Fortunately, there was no complaint, and he got away with his bluff!

22 The oldest profession

Three men after death turned up at the gates of heaven for entry. St Peter, who was supervising the entries to heaven, came up and said, "Sorry, we are nearly full up. There is only room for one. I will take the person who had the oldest profession."

The first man in the line declared that he was a surgeon. "Surely mine is the oldest profession. Recall that when Adam produced Eve from his ribs, that was as an act of surgery."

The second man came with another result. He said that he was a landscape artist. "Thus the profession of gardener is older because the Garden of Eden came up before Adam and Eve."

"What about you, sir?" asked St Peter to the third man.

He replied, "I was a cosmologist."

As soon as he said that, St Peter gave his verdict: "Cosmologists are known to create chaos in their work. Chaos preceded the Garden of Eden. So yours is the oldest profession."

23 When God was subjected to peer review

Once, in modern times, God decided to do some research on creation. As is the modern custom, He was advised to submit a research proposal to the concerned funding agency. The proposal was duly evaluated and eventually rejected. God was given feedback in the form of three reasons why his proposal was rejected.

1 It had been a long time since He last worked in the stipulated field.
2 No one else had been able to repeat his earlier work.
3 Although he had published his earlier work, he had not done so in a refereed high-profile journal but only in a book.

24 An International Astronomical Union (IAU) meeting in Baghdad

Almost exactly 1,200 years ago, Abdullah al Mansur, the second Abbasid Caliph, invited an international meeting of scientists and mathematicians to celebrate the founding of his new capital Baghdad. To this conference were invited scholars from all over, Greeks, Nestorians, Byzantines, Jews and Hindus. This was the first international meeting in an Arab country, and it probably initiated the renaissance of Islamic science. The main topic of the conference was observational astronomy, and Al Mansur asked the scholars present to prepare more accurate astronomical tables. He also ordered a more accurate determination of the circumference of Earth.

Looking back at this meeting, the most significant paper was read by Kankah, a delegate from India whose work revolutionized mathematical thinking and practice. The number system originating from India was made known internationally at this meeting.

25 Undisciplined white dwarfs

On 11 January 1935, the presentation of a paper at the Royal Astronomical Society by young scientist S. Chandrasekhar concluded with a limiting mass (about 1.4 solar masses) below which a white dwarf star can maintain equilibrium. The distinguished scientist Arthur Stanley Eddington did not like this conclusion and had this to say:

> Chandrassekhar shows that a star of mass greater than a certain limit M . . . has to go on radiating and radiating and contracting and contracting until, I suppose it gets to a few kilometers radius when gravity becomes strong enough to hold in the radiation, and the star can at last find peace. . . . I think there should be a law of nature to prevent a star from behaving in this absurd way.

These 'absurd' stars are today called black holes!

26 Tigers and theoreticians

Chip Arp, a distinguished observer was describing his work to a group of peers. He was presenting evidence that a quasar and a galaxy were near neighbours. As the two objects had different redshifts, many in the audience refused to accept Chip's data. Exasperated, he narrated a tale. Two astronomers were going through a forest when they saw a tiger coming.

"Run for your life," shouted one, and he quickly climbed a tree.

His colleague said, "My calculations show that we cannot have a tiger so near. I am sure that this tiger is distant. I will calculate and show you."

While he was calculating the answer, the tiger came and ate him.

27 How old is the universe?

Two Cambridge professors met outside the Cavendish Labs. One of them, Rutherford, asked the other, Eddington, "Professor Eddington, how old is the universe?"

Eddington did not expect that an experimental physicist would be used to long-time scales. To impress him, Eddington replied slowly, "My calculation shows that the universe is two thousand million years old."

Was Rutherford impressed? Hardly! He produced a piece of rock from his pocket and said, "My experiment shows that this piece is as old as three thousand million years."

28 The absent-minded mathematician

Norbert Wiener, a brilliant mathematician, was known to be absent-minded. His wife looked after all practical issues of the day since Norbert could not be trusted with any of them. The only matter he was trusted with was going to the department in the morning and returning home in the evening.

However, there was a change of routine one day because that day they were moving to a new home. As Norbert got into his car as usual, his wife gave him a piece of paper with instructions on how to get to his new house on his way back. He was specially told to follow them.

However, while at his office, he thought of a new idea, and to verify it, he looked for a paper in his pocket. He had that sheet of instructions given by his wife. However, he had forgotten what it was about and used it for writing the notes related to his idea. But after working on it, he found that the idea did not come up to his expectation, and he crumpled the paper and threw it in the waste-paper basket.

While coming back in the evening, he, of course, had forgotten about the change of house. As per his usual practice, he turned up at his old house to find it locked. He was puzzled and sought somebody around who could help. He saw a teenage girl on the porch whose face looked familiar.

He asked her, "I am Norbert Wiener who used to live here. What has happened to my family? How can I find them?"

The girl replied, "Daddy, mamma expected you to come to the old house, and in anticipation, she asked me to wait here. I will take you to our new house.

29 What do you see?

Three friends, an astronomer, a physicist and a mathematician were on a hiking tour near the English–Scottish border. At one vantage point, they found themselves in England looking across at Scotland. Over a distance of about 30 m, they saw a sheep grazing. Looking at the animal, the astronomer exclaimed, "So the sheep in Scotland are black."

The physicist was shocked by this observation and said, "You astronomers draw sweeping conclusions from scanty data. You need to take samples from different parts of Scotland, and if you find in a statistically significant way that sheep are black over there, only then can you make any such claim. What do you think?" He asked the mathematician for his view.

The mathematician shook his head and said, "You are both wrong! On the basis of what you see, all you can say is that the animal over there is black on the side facing us."

30 Euler and proof of God's existence

This story has not been historically confirmed but is nevertheless worth inclusion in this collection. It relates to the period Euler spent in Russia as a respected figure in the court of Empress Catherine. The French philosopher and atheist Diderot happened to visit Russia and called on the empress. He placed a challenge to the thinkers in Russia to argue with him on his premise of denying the existence of God.

Such was the reputation of Diderot that none of the local philosophers dared take up the challenge. As the court's reputation was at stake, the empress called Euler to come and argue with Diderot. Euler came, and when the debate began, he wrote on a blackboard the following mathematical formula:

$$\{a + b^n\}/n = X.$$

On the basis of which, he could prove that God exists.

Diderot knew no mathematics and was scared of mathematical formulae. What Euler had written was nonsense, but Diderot had no means of knowing so. Early the following day, he quietly left Russia and returned to Paris.

31 When maths and cricket combine

Eddington was a frequent visitor to the Cambridge University Fenner's cricket ground. While on one such visit, he formulated a cricket problem. The problem is too long to be given here. But basically, it demonstrates the logical structure of mathematics. Thus it gives the runs scored by one side, some bowling details of the other side, together with the condition that all runs were scored in ones and fours. Given these details, the reader is invited to give ball-by-ball details of what happened in the battings innings.

The problem is given in an appendix to Fred Hoyle's book *Nature of the Universe*.

32 Expectations from a mathematician

The old belief that a mathematician does number crunching is no longer borne out today. Rather, it is believed today that any type of mathematics is a work of art whose quality is judged by the quality of its basic assumptions and deductions. The notions involved in a particular branch of maths are judged by the quality, generality, disprovability of its assumptions, etc. These are abstract notions and nowhere do they involve the real world we live in.

As the philosopher and logician Bertrand Russell has remarked, a mathematician does not know what he is talking about, nor can he say as to what he is talking about is true. Of course, what he is talking about is logically consistent.

33 Telescope for the heaven

William Herschel was an efficient observer who appreciated the use of instruments in probing the universe. In particular, he had constructed two telescopes: one small but very efficient and the other large but not as fault-free as the smaller one. The king of England, George III, was a good patron for Herschel's observatory. He was quite impressed by Herschel's telescope and encouraged others to observe through them. Once he brought the archbishop of Canterbury for observing, saying, "My Lord, Bishop! You talk of heaven! Come, let me show you what it looks like."

This (larger) telescope had a 40-ft-long tube, which gave it its name 'the Great 40-feet telescope.' It was constructed by Herschel with the help of his sister Caroline who was herself a good observer. Completed in 1789, it was the largest telescope of its time and its cost of about 4,000 pounds was paid by the king.

34 Copernicus the healer

The same Copernicus who launched the heliocentric theory was also trained as a medical practitioner with training at the famous medical school at the University of Crakow, followed by a three-year medical course at the Padua University. The late Patrik Moor, well-known as an astronomy popularizer, describes in his book *Fireside Astronomy* a remedy claimed to be 'universal' recommended by Copernicus:

"Take two ounces of Armenian clay, a half ounce of cinnamon, two drachms of tormentil, dittany, red sandalwood, a drachm of ivory and iron shavings, two scruples of ash and rust, one drachm each of lemon peel and pearls; add one scruple each of emerald, red hyacinth and sapphire; one drachm of bone from a deer's heart; sea locusts, horn of a unicorn, red coral, gold and silver foil – all one scruple each; then add half a pound of sugar or the quantity which one buys for one Hungarian ducat's worth."

Then he adds a postscript: "God willing, it will help."

35 How useful is a baby?

Queen Victoria was visiting the famous laboratory of Michel Faraday. After witnessing various experiments on electric motors and dynamos, the sovereign was suitably impressed. But she was concerned about one aspect. She asked the scientist, "Sir, what you showed me is very much full of interest; but may I ask, what is the use of all these experiments?"

Faraday dared to make a counter remark, "Sire! Do you ask what is the use of a newborn baby?"

***

(In another version of this story, the prime minister was the visitor who asked the same question, and Faraday replied: "Sir, one day you will be able to tax the products made from these experiments.")

36 What is my mother tongue?

A diplomat who could speak fluently many languages came to the court of the Pehwas (the then rulers of southern India). He challenged those present to figure out his mother tongue. This turned out to be nearly impossible, as he spoke all languages with equal facility. At last, the Peshwa called upon Nana Phadnis, his main administrator, to take up the challenge.

Nana invited the visitor to be his house guest for the night since it was already getting late. Nana was a good host, and the guest retired for the night relaxed and well-fed. However, during the night, a shocking experience was in store for him. Someone poured a large jugful of ice-cold water on the sleeping guest. While the guest was muttering angrily, Nana turned up to apologize. He promised suitable punishment to whosoever was responsible. He offered a new bed and dry clothes so that the guest was comfortable.

The next morning, they all met in the court, and Nana confidently asserted that the guest's mother tongue was Kannada. The answer was correct, and the guest asked Nana how he found out.

"That was the language you used in your instantaneous reaction to the shock treatment you had when cold water was poured on you, sir."

Albert Einstein spoke English well and could argue with critics in that language. However, when he got deeply involved in arguments and wished to make a strong point, he would relapse into German, his mother tongue.

INDEX

Printed in the United States
by Baker & Taylor Publisher Services

Printed in the United States
by Baker & Taylor Publisher Services